Adsorption From Aqueous Solution

Adsorption From Aqueous Solution

A symposium co-sponsored by the Division of Water, Air, and Waste Chemistry and the Division of Colloid and Surface Chemistry at the 154th Meeting of the American Chemical Society Chicago, Illinois, Sept. 14-15, 1967.

Walter J. Weber, Jr. and Egon Matijević
Symposium Chairmen

ADVANCES IN CHEMISTRY SERIES **79**

AMERICAN CHEMICAL SOCIETY
WASHINGTON, D. C.   1968

Copyright © 1968

American Chemical Society

All Rights Reserved

Library of Congress Catalog Card 68–59407

PRINTED IN THE UNITED STATES OF AMERICA

Advances in Chemistry Series

Robert F. Gould, *Editor*

Advisory Board

Sidney M. Cantor

Frank G. Ciapetta

William von Fischer

Edward L. Haenisch

Edwin J. Hart

Stanley Kirschner

John L. Lundberg

Harry S. Mosher

Edward E. Smissman

AMERICAN CHEMICAL SOCIETY PUBLICATIONS

FOREWORD

ADVANCES IN CHEMISTRY SERIES was founded in 1949 by the American Chemical Society as an outlet for symposia and collections of data in special areas of topical interest that could not be accommodated in the Society's journals. It provides a medium for symposia that would otherwise be fragmented, their papers distributed among several journals or not published at all. Papers are refereed critically according to ACS editorial standards and receive the careful attention and processing characteristic of ACS publications. Papers published in ADVANCES IN CHEMISTRY SERIES are original contributions not published elsewhere in whole or major part and include reports of research as well as reviews since symposia may embrace both types of presentation.

CONTENTS

Preface ix

1. **Capillary Thermodynamics Without a Geometric Gibbs Convention** ... 1
   F. C. Goodrich, Institute of Colloid and Surface Science, Clarkson College of Technology, Potsdam, N. Y.

2. **Kinetics of Adsorption** 8
   J. M. Smith, University of California, Davis, Calif.

3. **Two-Dimensional Monomolecular Ion Exchangers. Kinetics and Equilibrium of Ion Exchange** 23
   M. de Heaulme, Y. Hendrikx, A. Luzzati, and L. Ter Minassian-Saraga, Centre National de la Recherche Scientifique, Physico-Chimie des Membranes, 45, rue des Saints-Peres, Paris VI°, France

4. **Chromatographic Behavior of Interfering Solutes. Conceptual Basis and Outline of a General Theory** 30
   F. Helfferich, Shell Development Company, Emeryville, Calif.

5. **Adsorption of Hydrolyzed Hafnium Ions on Glass** 44
   Lynden J. Stryker and Egon Matijević, Department of Chemistry and Institute of Colloid and Surface Science, Clarkson College of Technology, Potsdam, N. Y.

6. **The Adsorption of Aqueous Co(II) at the Silica-Water Interface** .. 62
   T. W. Healy, R. O. James, and R. Cooper, Department of Physical Chemistry, University of Melbourne, Parkville, Victoria 3052, Australia

7. **The Adsorption of Aqueous Metal on Colloidal Hydrous Manganese Oxide** 74
   D. J. Murray, Department of Mineral Technology, University of California, Berkeley, Calif., T. W. Healy, Department of Chemistry, University of Melbourne, Victoria, Australia, and D. W. Fuerstenau, Department of Mineral Technology, College of Engineering, University of California, Berkeley, Calif.

8. **Adsorption of Selenite by Goethite** 82
   F. J. Hingston, Division of Soils, C.S.I.R.O., W. A., Laboratories, Wembley, Western Australia, 6014, and A. M. Posner and J. P. Quirk, Department of Soil Science and Plant Nutrition, University of Western Australia, Nedlands, Western Australia, 6009

9. Coagulation by Al(III). The Role of Adsorption of Hydrolyzed Aluminum in the Kinetics of Coagulation 91
   Hermann H. Hahn and Werner Stumm, Division of Engineering and Applied Physics, Harvard University, Cambridge, Mass.

10. Reaction of the Hydrated Proton with Active Carbon 112
    Vernon L. Snoeyink and Walter J. Weber, Jr., Department of Civil Engineering, The University of Michigan, Ann Arbor, Mich.

11. Adsorption and Wetting Phenomena Associated with Graphon in Aqueous Surfactant Solutions 135
    F. G. Greenwood, G. D. Parfitt, N. H. Picton, and D. G. Wharton, University of Nottingham, Nottingham, England

12. Analysis of the Composite Isotherm for the Adsorption of a Strong Electrolyte from Its Aqueous Solution onto a Solid 145
    Karma M. Van Dolsen and Marjorie J. Vold, The University of Southern California, Los Angeles, Calif.

13. The Effect of Hydrocarbon Chain Length on the Adsorption of Sulfonates at the Solid/Water Interface 161
    T. Wakamatsu and D. W. Fuerstenau, College of Engineering, University of California, Berkeley, Calif.

14. Adsorption of Dyes and Their Surface Spectra 173
    A. H. Herz, R. P. Danner, and G. A. Janusonis, Research Laboratories, Eastman Kodak Company, Rochester, N. Y.

15. The Relative Adsorbability of Counterions at the Charged Interface   198
    Kozo Shinoda and Masamichi Fujihira, Yokohama National University, Ookamachi, Minamiku, Yokohama, Japan

Index 207

PREFACE

The adsorption of soluble species at liquid-solid interfaces has been of long standing interest to chemists, biologists, and engineers concerned with the characteristics and phenomenological behavior of a variety of sorbate-sorbent systems. Analytical chemists recognized very early that sorption reactions can have significant effects on the results of even simple gravimetric analyses. Colloid chemists near the turn of the century established the critical relationship between the stability of lyophobic sols and the presence of electrolytes in solution phase. For some time, coagulation by electrolytes was thought to be effected essentially by charge neutralization resulting from counterion adsorption. Although this hypothesis has been disproved, it is widely recognized that the adsorption of counterions does have a significant role in the formation of the electrical double layer. Further, specific adsorption of complex ions at liquid-solid interfaces plays a predominant role in colloid stability and in such interfacial processes as adhesion, flotation, heterogeneous catalysis, etc.

Adsorption has for some time been recognized in the chemical industries as an effective process for a wide range of solute-solvent separations. The past several years have witnessed a rapidly mounting interest in adsorption as a unit operation for water and waste water treatment. Beyond this, there is an increasing realization of the significance of adsorption phenomena in a large number and wide variety of other treatment operations and natural processes associated with water.

Current interest in the multi-faceted role of adsorption in the water and waste water field has been stimulated in large part by research originating within this field. Concomitantly, more and more investigations in the field of surface and colloid chemistry have focused on aqueous systems. The results of these investigations are of considerable significance to the water scientist. The convergence of interests of researchers in these two fields has provided the impetus for this joint symposium on adsorption from aqueous solution.

In any consideration of what is presently known regarding adsorption from aqueous solution it is necessary to distinguish between neutral species, simple ions, complex ions, surface active agents, polyelectrolytes, and charged and uncharged high polymers. Much of the data available in the literature relates to the adsorption of polymers and complex-forming ions, with few reports on the adsorption of neutral species. Most discussion relates to equilibrium or quasi-equilibrium conditions. Recently the importance of the kinetics of adsorption from aqueous solution has become increasingly evident. Additionally, certain types of sorption phenomena—such as the adsorption of charged polymers on colloidal particles of like charge—are not at all well understood. It is apparent that, despite a rather abundant general literature on adsorption, much remains to be answered regarding the accumulation of various species at solid-liquid interfaces in aqueous systems.

Several pertinent properties of a solute species must be clearly defined for a proper understanding of its adsorptivity in a given aqueous system. These include: (a) its charge characteristics; (b) its size and shape; (c) the ratio of its polar and nonpolar segments; (d) the chemical properties of its ligands in the case of a complex species; and (e) its orientation in the interface of interest. Adsorptivity in a given system will also depend greatly on the surface and chemical properties of the adsorbent, on the number and structural arrangement of water molecules at the surface, and on the chemical composition of the bulk aqueous phase.

The papers in this volume deal with many of the foregoing questions and problems relating to adsorption from aqueous solution. In addition to general discussions of thermodynamic and kinetic aspects of adsorption phenomena, the papers include description of the results of studies on a variety of adsorbate-adsorbent systems. Among the adsorbates studied are: (1) strong electrolytes; (2) unhydrolyzed multi-valent cations; (3) hydrolyzed metal ions; (4) complex ions; (5) organic dyes; (6) organic pesticides; (7) surface active agents; and (8) macro-molecules. Adsorbents studied include: (1) glass; (2) silica; (3) silver halides; (4) rutile; (5) manganese dioxide; (6) alumina; (7) goethite; (8) arsenic sulfide; (9) graphon; and, (10) active carbon. Several authors have described the results of studies on ion exchange as a special aspect of adsorption.

The papers contained in this volume, although leaving many questions unanswered and many problems unsolved, should prove a significant contribution to a general understanding of processes involved in adsorption from aqueous solution.

The symposium has been supported in part by Grant WP 01132-01 from the Federal Water Pollution Control Administration, U.S. Department of Interior.

Ann Arbor, Michigan
Potsdam, New York
June 1968

WALTER J. WEBER, JR.
EGON MATIJEVIĆ

Capillary Thermodynamics Without a Geometric Gibbs Convention

F. C. GOODRICH

Institute of Colloid and Surface Science, Clarkson College of Technology, Potsdam, N. Y.

> *It is shown in a number of important special cases that the geometric conventions employed by Gibbs in his treatment of capillary thermodynamics are replaceable by an algebraic formalism in which no mention is made of "dividing surfaces." The resulting capillary excess quantities are therefore explicitly defined and may be interrelated by standard algebraic operations. To the extent that algebra is a more reliable tool than the intuitive manipulation of geometric surfaces, the method is preferable to the original development of Gibbs.*

Gibbs' thermodynamic analysis of fluid systems containing a plane interface (2) is characterized by the construction through the interfacial region of imaginary mathematical surfaces which are supposed to locate the extent of the separate phases in an equivalent model system in which by convention everything to one side of the mathematical surface is homogeneous phase α while everything to the other side is homogeneous phase β. Gibbs well understood the arbitrariness of this division, and indeed in his work made use of this arbitrariness to draw different mathematical surfaces through the interfacial region, each such mathematical surface being involved with a different set of thermodynamic quantities associated with the capillary layer.

Whatever may be said for the mathematical utility of this procedure, its pedagogic results have been disastrous, and I know of no other widely used area of thermodynamics where so much confusion exists. Even to this day the literature is full of errors and imprecisely defined capillary quantities, all of them going back ultimately to a too literal interpretation of the Gibbs dividing surfaces. Guggenheim (3) and Defay (1) have partly alleviated this situation by emphasizing that experimentally measurable quantities must in the thermodynamic equations be invariant to the location of a dividing surface, but as has been shown in a special case by Hansen (4), it is both possible and preferable to handle the thermodynamics without ever introducing mathematical surfaces at all, and this will be the approach adopted here.

One Component Systems

As a first example, consider a pure liquid in equilibrium with its vapor. Because I wish to focus attention on the liquid/gas interface to the exclusion of adsorption effects at solid boundaries, I shall suppose the containing vessel to be chemically inert. The Gibbs-Duhem equation for the system is then

$$SdT - VdP + Ad\gamma + nd\mu = 0, \tag{1}$$

in which the symbols have their conventional meanings and where the extensive quantities S, V, A, and n refer to the entire two phase system enclosed by the containing vessel and not to some part of it isolated artificially from the rest by mathematical boundaries. It is not uncommon in the literature to find derived from (1) equations of the type

$$(\partial \gamma / \partial T)_{P\mu} = -S/A \tag{2}$$

which are completely fallacious, for a one component, two phase system has from the phase rule only one degree of freedom, so that it is impossible to alter the temperature of the system without simultaneously altering the pressure and the chemical potential. This is to say that Equation 2 is false because the intensive variables T, P, γ, and μ are not independently variable.

The latter fact may be emphasized by considering small samples of molar content n^α and n^β drawn from the bulk phases in regions far from the interface. The size and shape of these samples need have no relationship to the geometry of the interface—any irregularly shaped specimen of bulk phase will do. For each of these bulk phase samples α and β we have Gibbs-Duhem equations

$$S^\alpha dT - V^\alpha dP + n^\alpha d\mu = 0$$
$$S^\beta dT - V^\beta dP + n^\beta d\mu = 0$$

or upon dividing through each by n^α, n^β respectively

$$\begin{aligned}\overline{S}^\alpha dT - \overline{V}^\alpha dP + d\mu &= 0 \\ \overline{S}^\beta dT - \overline{V}^\beta dP + d\mu &= 0\end{aligned} \tag{3}$$

in which as usual the barred symbols are the molar entropies and volumes of the respective phases.

Equations 1 and 3 comprise a set of three relationships between four differentials, dT, dP, $d\gamma$, and $d\mu$. There thus remains a single degree of freedom in the system in agreement with the phase rule.

In order to eliminate unwanted variables, multiply Equations 3 by algebraic multipliers x and y respectively and subtract from Equation 1.

$$(S - x\bar{S}^\alpha - y\bar{S}^\beta)dT - (V - x\bar{V}^\alpha - y\bar{V}^\beta)dP + Ad\gamma + (n - x - y)d\mu = 0 \quad (4)$$

Equation 4 is valid for arbitrary choice of x and y, and we shall obtain different special equations for different choices of the multipliers. Each such choice corresponds to a different Gibbs convention. To understand this fact, note for example that a quantity $S - x\bar{S}^\alpha - y\bar{S}^\beta$ is the total entropy of the heterogeneous system minus an amount $x\bar{S}^\alpha$ of the entropy of the bulk α phase minus an amount $y\bar{S}^\beta$ of the bulk β phase. Depending therefore on how we choose x and y, we shall be defining excess entropies which compare the real, two phase system containing an interface with fictitious systems comprising two bulk phases of entropy $x\bar{S}^\alpha$ and $y\bar{S}^\beta$ in the absence of an interfacial region. The multipliers thus perform the same function as a Gibbs dividing surface, but because we are using algebraic methods instead of geometric intuition, we shall be able to make our thermodynamics more explicit.

To obtain an interpretation of the experimentally measurable quantity $d\gamma/dT$, choose x and y in Equation 4 to make the coefficients of dP and $d\mu$ vanish.

$$x + y = n$$
$$x\bar{V}^\alpha + y\bar{V}^\beta = V \quad (5)$$

These equations are solvable explicitly,

$$x = \frac{n\bar{V}^\beta - V}{\bar{V}^\beta - \bar{V}^\alpha} \; ; \; y = \frac{V - n\bar{V}^\alpha}{\bar{V}^\beta - \bar{V}^\alpha} \quad (6)$$

whence upon substitution into Equation 4 and rearrangement,

$$d\gamma/dT = -S^{(n,V)} \equiv -A^{-1}\left[S - \frac{n\bar{V}^\beta - V}{\bar{V}^\beta - \bar{V}^\alpha}\bar{S}^\alpha - \frac{V - n\bar{V}^\alpha}{\bar{V}^\beta - \bar{V}^\alpha}\bar{S}^\beta\right]. \quad (7)$$

The introduction of the superscript (n, V) implies the adoption of a Gibbs convention algebraically expressed by Equations 5, which state that we are comparing the real system with a fictitious one consisting of two bulk phases in contact in the absence of an interfacial region, and with the added specification that the numbers of moles n and the total volume V of the fictitious system shall be the same as in the real one. The quantity $S^{(n,V)}$ (which is called $S_A{}^\sigma$ by Guggenheim (3) and $s_1{}^\sigma$ by Defay (1)) is the entropy change per unit area of interface created when a tall, slender, closed vessel of fixed volume V containing n moles of a substance in gas-liquid equilibrium is laid isothermally on its side, thereby enlarging the interfacial area. Neither the total volume nor the total moles within the vessel change, and the entropy change is exclusively due to the creation of new interface from material previously present in the bulk phases.

Confidence in the correctness of Equation 7 is increased by an heuristic argument which will not be reproduced here, but which shows that x and y in Equations 6 are good estimates of the total numbers of moles of liquid and of vapor in the real system. It follows that $x\bar{S}^\alpha$ and $y\bar{S}^\beta$ are correspondingly good estimates of the total entropies to be assigned to the liquid and vapor phases in the real system, so that $S - x\bar{S}^\alpha - x\bar{S}^\beta$ is the amount by which the entropy of the real system exceeds the combined entropies of its liquid and vapor phases. The phrase "good estimate" here has more of a literary than a mathematical interpretation; for the diffuseness of the interfacial region prohibits any mathematically rigorous definition of the volumes of the respective phases, and this fact was the whole motivation of Gibbs' use of a dividing surface.

Two Component Systems

Turning now to adsorption equilibrium, let us apply algebraic methods to a two component 1,2 phase system. From the phase rule there will be two degrees of freedom, but we shall reduce this to one by maintaining the temperature constant. Then for the total system there exists a Gibbs-Duhem equation

$$-VdP + Ad\gamma + n_1 d\mu_1 + n_2 d\mu_2 = 0 \qquad (8)$$

and for samples of arbitrary shape and size drawn from the interiors of the bulk phases

$$\begin{aligned}-V^\alpha dP + n_1{}^\alpha d\mu_1 + n_2{}^\alpha d\mu_2 &= 0 \\ -V^\beta dP + n_1{}^\beta d\mu_1 + n_2{}^\beta d\mu_2 &= 0.\end{aligned} \qquad (9)$$

Dividing through Equations 9 by V^α and V^β respectively we obtain equations which are independent of the size of the samples chosen:

$$\begin{aligned}-dP + c_1{}^\alpha d\mu_1 + c_2{}^\alpha d\mu_2 &= 0 \\ -dP + c_1{}^\beta d\mu_1 + c_2{}^\beta d\mu_2 &= 0\end{aligned} \qquad (10)$$

in which the c's are the bulk phase concentrations in moles per liter of the several components in the indicated phases.

As in our previous work, Equations 8 and 10 comprise three relations between four differentials, so that in agreement with the phase rule a single relation between two differentials is implied. To obtain such a relation, multiply Equations 10 by multipliers x and y and subtract from Equation 8.

$$-(V - x - y)dP + Ad\gamma + (n_1 - xc_1^\alpha - yc_1^\beta)d\mu_1 + (n_2 - xc_2^\alpha - yc_2^\beta)d\mu_2 = 0 \quad (11)$$

We have conceptually arrived at the same point as we did in the argument leading to Equation 4; for Equation 11 is valid for arbitrary choice of multipliers x and y, and each such choice corresponds to a different Gibbs convention. The choice which leads to the Gibbs adsorption equation is that which makes the coefficients of dP and $d\mu_1$ (conventionally defined to be the solvent) vanish:

$$x + y = V$$
$$xc_1^\alpha + yc_1^\beta = n_1 \quad (12)$$

which is to say

$$x = \frac{n_1 - Vc_1^\beta}{c_1^\alpha - c_1^\beta} \quad ; \quad y = \frac{Vc_1^\alpha - n_1}{c_1^\alpha - c_1^\beta}. \quad (13)$$

Substituting into Equation 11 and rearranging,

$$(\partial \gamma / \partial \mu_2)_T = -\Gamma_2^{(1)} \equiv -A^{-1}\left[n_2 - \frac{n_1 - Vc_1^\alpha}{c_1^\alpha - c_1^\beta}c_2^\alpha - \frac{Vc_1^\alpha - n_1}{c_1^\alpha - c_1^\beta}c_2^\beta\right]. \quad (14)$$

Two things are to be noted about Equation 14. The first is that the pressure is not held constant. I emphasize this point because the statement is frequently seen in the literature that the Gibbs adsorption equation is valid only under conditions of constant temperature and pressure, a restriction which for a two phase, two component system reduces the number of degrees of freedom to zero. The second thing to note is that the Gibbs convention implied by Equation 14 is contained in Equations 12, which state that we are comparing the real system with a fictitious one having the same volume and the same numbers of moles of solvent as the real system. Finally an heuristic argument valid for most conditions of experimental interest shows that the quantity in brackets on the right hand side of Equation 14 is the difference between the total moles n_2 of solute actually present in the system minus the total moles of solute present in the bulk liquid and vapor phases.

Another excess quantity $\Gamma_1^{(2)}$ is defined by interchanging indices 1 and 2 in Equations 12, 13, and 14, and for this quantity the Gibbs convention is different, for we are now comparing the real system with a fictitious one defined to have the same total moles n_2 of solute as has the real one. Despite the fact that $\Gamma_1^{(2)}$ and $\Gamma_2^{(1)}$ are defined for different Gibbs conventions, they are algebraically related, as the reader may readily confirm for himself. Directly from their algebraic definitions one can show that

$$(c_1^\alpha - c_1^\beta)\Gamma_2^{(1)} + (c_2^\alpha - c_2^\beta)\Gamma_1^{(2)} = 0. \quad (15)$$

This equation assumes a more familiar form if we are willing to ignore the gas phase (β) concentrations in comparison with those of the liquid phase, for then approximately

$$c_1^\alpha \Gamma_2^{(1)} + c_2^\alpha \Gamma_1^{(2)} = 0,$$

or what is the same thing

$$x_1 \Gamma_2^{(1)} + x_2 \Gamma_1^{(2)} = 0, \quad (16)$$

in which x_1 and x_2 are the mole fractions of the liquid phase. Equation 16 is familiar, being quoted, for example, by Defay (*1*).

As a final example, let us derive another set of excess quantities from Equation 11 by choosing the Gibbs convention

$$x + y = V$$
$$x(c_1^\alpha + c_2^\alpha) + y(c_1^\beta + c_2^\beta) = (n_1 + n_2). \quad (17)$$

Upon solving these equations for x and y and introducing the results into Equation 11 we have an adsorption equation

$$d\gamma + \Gamma_1^{(n)} d\mu_1 + \Gamma_2^{(n)} d\mu_2 = 0 \quad (18)$$

in which

$$\Gamma_1^{(n)} \equiv A^{-1} \left[n_1 - \frac{(n_1 + n_2) - V(c_1^\beta + c_2^\beta)}{(c_1^\alpha + c_2^\alpha) - (c_1^\beta + c_2^\beta)} c_1^\alpha - \frac{V(c_1^\alpha + c_2^\alpha) - (n_1 + n_2)}{(c_1^\alpha + c_2^\alpha) - (c_1^\beta + c_2^\beta)} c_1^\beta \right] \quad (19)$$

and a similar equation for $\Gamma_2^{(n)}$ derived from Equation 19 by interchanging indices 1,2. While the appearance of Equation 19 is formidable, the quantities $\Gamma_1^{(n)}$ and $\Gamma_2^{(n)}$ are very precisely defined and may be shown directly from their algebraic definitions to satisfy the linear relation

$$\Gamma_1^{(n)} + \Gamma_2^{(n)} = 0 \quad (20)$$

which is to be compared with Equation 15.

The Gibbs convention (Equation 17) states that we compare the real system with a fictitious one having the same total volume and total numbers of moles of all constituents as the real system. Under this convention the total moles of individual components 1 and 2 will differ between the real and the fictitious systems, but because the total of all moles of both components is the same, the surface excesses of each must sum to zero, and this is the meaning of Equation 20.

This convention has the advantage that both of the consituents of the mixture are defined according to the same convention, whereas the quantities $\Gamma_1^{(2)}$ and $\Gamma_2^{(1)}$ are separately defined by different conventions. It would be a matter of some difficulty to prove by geometric manipulation of dividing surfaces that, for example, $\Gamma_2^{(1)}$ and $\Gamma_2^{(n)}$ are linearly related. From their algebraic definitions, however, it is easy to show that

$$\Gamma_2^{(1)} = \left[1 + \frac{c_2^\alpha - c_2^\beta}{c_1^\alpha - c_1^\beta}\right] \Gamma_2^{(n)} \qquad (21)$$

exactly. If we neglect the concentrations in the gas (β) phase, this simplifies to

$$\Gamma_2^{(n)} = x_1 \Gamma_2^{(1)}, \qquad (22)$$

which is the relation most commonly found in the literature (5, 6).

While I have given here only a few examples, it turns out that the entirety of capillary thermodynamics may be founded upon algebraic methods, resulting in a great improvement in the ease with which the capillary excess quantities may be defined and manipulated.

Acknowledgment

I am indebted to G. Schay for showing me a copy of his article in advance of publication.

Literature Cited

(1) Defay, R., Priogogine, I., Bellemans, A., Everett, D. H., "Surface Tension and Adsorption," Wiley and Sons, Inc., New York (1966).
(2) Gibbs, J. W., "Collected Works," Vol. I, Yale University Press, New Haven (1948).
(3) Guggenheim, E. A., "Thermodynamics," Interscience, New York (1949).
(4) Hansen, R. S., *J. Phys. Chem.* **66**, 410 (1962).
(5) Kipling, J. J., "Adsorption from Solutions of Non-Electrolytes," Academic Press, New York (1965).
(6) Schay, G., "Surface and Colloid Science," Interscience, New York (in press).

RECEIVED October 26, 1967.

2

Kinetics of Adsorption

J. M. SMITH
University of California, Davis, Calif.

> *The effects of physical transport processes on the overall adsorption on porous solids are discussed. Quantitative models are presented by which these effects can be taken into account in designing adsorption equipment or in interpreting observed data. Intraparticle processes are often of major importance in adsorption kinetics, particularly for liquid systems. The diffusivities which describe intraparticle transfer are complex, even for gaseous adsorbates. More than a single rate coefficient is commonly necessary to represent correctly the mass transfer in the interior of the adsorbent.*

Adsorption is normally thought of as the process by which a molecule or atom in a fluid is attached to a solid surface, and it is implied that the molecule (or atom) is in the same location as the site. Kinetics of such processes is concerned with force fields between sites and molecules and forms an important area of surface chemistry. However, in this paper both a wider and more restricted view will be taken of adsorption kinetics in that emphasis will be put on the so-called physical processes that must accompany adsorption, if the overall process is to continue. In particular the kind of kinetics discussed will be that necessary to explain the performance of, or to design an apparatus for, separating or removing components in a fluid stream.

The use of the term physical to distinguish other steps from the process at the site, while common, is unfortunate. The so-called physical processes of diffusion may involve mechanisms best described by chemical means. For example, surface diffusion on the pore walls is an activated process. On the other hand, the adsorption itself at a surface site may involve such weak forces that a physical rather than chemical bond best characterizes the mechanics. What is sought is a distinction between the adsorption step at a fixed location and processes by which the adsorbent is transferred to the site. The distinguishing feature is whether the process involves the movement of the adsorbate. Hence, a more correct characterization would be space or point processes. Applications of adsorption require transfer of the adsorbate between a bulk fluid and a site on the solid surface. Hence, the space processes, of necessity, do occur. A useful geometric arrangement for adsorption is an assembly of solid particles, permeable to the adsorbate, with the adsorbate-containing fluid flowing through the interparticle void space—the conventional fixed-bed absorber. Basic features of both types of processes have been taken into account in using this form of adsorption apparatus. Recognition that the rate depends upon the number of sites has necessitated permeable particles so that interior sites are potentially usable. Realization that the transfer distance must not be excessive from bulk fluid to site, or interior sites will be valueless, lead to the use of small particles. As intimated, the important point is not whether space processes are involved, but whether they affect the rate of adsorption. The results of adsorption are measured in terms of concentrations of the adsorbate in the bulk fluid; concentrations on the solid surfaces are not useful in the design or performance evaluation tasks. Such concentrations are usually unknown. What is needed are rate equations which express the kinetics of adsorption in terms of properties of bulk fluid adjacent to the solid particle—space or global kinetics in contrast to point kinetics, which express the rate in terms of properties at the adsorption site. Space processes are significant to the extent that variables such as fluid velocity and particle size affect the global kinetics equations.

From this point of view the proper approach to design of adsorption equipment is a two-step procedure:

A. Establish the point kinetics and equilibrium isotherm for the adsorption step at the site, and the rate coefficients for the various space processes. These are the two types of data required for design. To some extent estimates may be made for the coefficients for the space processes.

B. By applying the conservation equations (mass, energy) and the data from A, calculate the concentration as a function of time and position in the adsorption apparatus.

It is desirable to list the sequence of space and point steps which together constitute global adsorption. This is not a new concept and such descriptions have frequently been presented (1, 14), particularly for fluid reactions on porous catalyst particles. The first space process, axial dispersion, is not a part of the sequence, but it does affect the observed kinetics, and is logically considered as a space process. Its significance depends upon the reciprocal of the axial Peclet number, $E_x/(2R)v$. The sequential steps are:

(1) External transfer: transfer of the absorbate from bulk fluid to outer surface of particle by molecular and convective diffusion.

(2) Internal transfer: transfer of adsorbate from particle surface to interior site; by diffusion in the void space of the pores, by surface migration on the pore surface, or by volume diffusion, for example, in the holes in the chemical structure of the solid phase.

(3) Adsorption on the interior site. For aqueous solutions such adsorption is usually classified by either ionic type chemical steps, or by a relatively loose adsorption with low heat of adsorption, that is by physical adsorption. In either case the process has a high rate constant so that this step usually does not influence the global kinetics.

(4) In replacement or reaction operations where a product is produced, the same steps as (2) and (3) occur for the product in the reverse direction.

While the treatment so far has not considered the effect of temperature on global kinetics, heats of adsorption or chemical reaction can be significant. Temperature gradients are generally less for liquid systems —e.g., aqueous solutions—than for gaseous since the heat capacity ($c_p \rho$) of the liquid stream is an order of magnitude higher than that for gases. However, to present a more complete approach to the problem, the steps by which the heat of adsorption is transferred to the flowing stream are:

1. Heat release caused by adsorption at the interior site.

2. Internal transfer: transfer of energy to the outer surface of the solid particle. This is commonly treated as though the particle was homogeneous with a single effective thermal conductivity.

3. External transfer: transfer of energy from the surface of the particle into the fluid stream. The properties of flowing fluids are such that the resistance to heat transfer can be larger than that for mass transfer, so that a negligible concentration difference may exist between bulk fluid and particle surface and yet the corresponding temperature difference will be significant.

4. Axial dispersion of energy along the fluid stream. The importance of this process is determined by the reciprocal of the Peclet number for heat transfer, $k_x/(2R)\rho C_f$.

Conservation Equations

If the processes just described are assumed to characterize the transfer of mass and energy in a fixed-bed adsorber, the conservation principles may be applied to them to describe the temperature and concentration as a function of time and position. Presenting the equations for a fixed-bed geometry has the advantage of including also equations, as special cases, for transient adsorption in single particles or groups of particles in batch systems.

In aqueous systems, as in others, some of the steps in the transfer processes have relatively high rate coefficients. Such steps may be treated as occurring at near equilibrium so that the driving force, either a concentration or temperature difference, approaches zero. The simplification in the treatment under those conditions can be readily seen from the equations. Let us suppose that a single species is adsorbed from an otherwise inert stream flowing through a fixed bed of porous particles (external porosity = α, internal porosity = β). The internal diffusivity D_c is based upon the total area perpendicular to the radial direction in the particle so that for non-porous particles the only change would be $\beta = 1$. The equations are written for spherical particles. A number of assumptions are made, such as diffusion only in the radial direction, but these are evident from the form of the equations and need not be described in detail. It should be mentioned that a linear form is assumed for the point adsorption kinetics (Equation 4). Doing this results in linear equations in concentration which may be solved by standard mathematical methods. Equation 4 is a simplified form of the Langmuir expression, applicable for gaseous systems when the extent of adsorption is low. For ion exchange in aqueous systems Equation 4 would correspond to first order, reversible kinetics at the adsorption site.

Conservation of Mass. If c represents the concentration of adsorbent in the bulk fluid stream, c_i the intraparticle concentration, and c_{ads} the concentration of adsorbed component, mass conservation in the fluid phase requires

$$\frac{E_x}{\alpha} \frac{\delta^2 c}{\delta z^2} - v \frac{\delta c}{\delta z} - \frac{\delta c}{\delta t} - R_v = 0 \qquad (1)$$

and within the particle

$$\frac{D_c}{\beta} \frac{\delta^2 c_i}{\delta r^2} + \frac{2}{r} \frac{\delta c_i}{\delta r} - \frac{\delta c_i}{\delta t} - \frac{\rho_p}{\beta} \frac{\delta c_{ads}}{\delta t} = 0 \qquad (2)$$

The last term in Equation 1 represents the global kinetics, expressed as the rate of adsorption per unit volume of bed. It cannot be written in terms of concentrations in the fluid phase—the equations must be solved to do this—but it may be expressed quantitatively in terms of c_i by writing an expression for the diffusion rate:

$$R_v = \left(\frac{3D_c}{R}\right)\left(\frac{1-\alpha}{\alpha}\right)\left(\frac{\delta c_i}{\delta r}\right)_{r=R} \qquad (3)$$

The point kinetics, in terms of a first order, reversible adsorption rate is

$$R_p = \rho_p \frac{\delta c_{ads}}{\delta t} = k_{ads} \rho_p (c_i - c_{ads}/K) \qquad (4)$$

If other kinetics represent the point process, the problem is still defined as long as R_p can be written in terms of concentrations at the adsorption site, although the solution will be more difficult.

The external and internal concentrations are related by Step 1 of the sequence:

$$D_c \left(\frac{\delta c_i}{\delta r}\right)_{r=R} = k_f(c - c_i) \quad (5)$$

Return now to the first part of the two-step procedure for characterizing an adsorption column. Our model represented by Equations 1, 2, 3, 4, and 5 shows that the following space and point rate coefficients define the problem:

Space coefficients: E_x, k_f, D_c

Point coefficient: k_{ads}

Reliable values for all these coefficients would be required for situations where all the steps are significant.

Conservation of Energy. The conservation equation for energy in the fluid phase is similar to Equation 1:

$$\frac{k_x}{\alpha} \frac{\delta^2 T}{\delta z^2} - \rho_f C_f \left(v \frac{\delta T}{\delta z} - \frac{\delta T}{\delta t} \right) - Q_v = 0 \quad (6)$$

where Q_v is the rate of heat transfer from particle to fluid phase per unit volume; that is

$$Q_v = \frac{3k_e}{R} \left(\frac{1-\alpha}{\alpha}\right) \left(\frac{\delta T_i}{\delta r}\right)_{r=R} \quad (7)$$

The energy balance within the particle includes the heat of adsorption, Q_p per unit volume of particle; that is,

$$k_e \left(\frac{\delta^2 T_i}{\delta r^2} + \frac{2}{r} \frac{\delta T_i}{\delta r} \right) - \rho_p C_p \frac{\delta T_i}{\delta t} - Q_p = 0 \quad (8)$$

where

$$Q_p = (-\Delta H) R_p = (-\Delta H) k_{ads} \rho_p (c_i - c_{ads}/K) \quad (9)$$

The equation relating T_i and T is analogous to Equation 5, or

$$k_e \left(\frac{\delta T_i}{\delta r}\right)_{r=R} = h_f(T - T_i) \quad (10)$$

Temperature has a strong influence on k_{ads} and sometimes on D_c. If the specific step described by these two coefficients has a significant effect on the kinetics of adsorption, then energy conservation equations may have to be included in the analysis to establish the temperature to every point in the bed. In these circumstances the additional space coefficients, k_x, h_f, and k_e must be added to the previous list. A point coefficient analogous to k_{ads} is not included, because it has been assumed that the temperature behavior of the particle can be faithfully represented by assuming a homogenous material. With this simplification it is not necessary to introduce different temperatures within the particle for solid and porous regions.

To illustrate the nature of temperature effects for aqueous systems, consider an ion exchange bed. If the external diffusion step controls the global kinetics, for example as described by Schlogl and Helfferich (*12*), small temperature differences in the bed are not likely to be significant, since neither k_{ads} or D_c are necessary to describe the global kinetics. On the other hand if internal diffusion has an effect on the rate of adsorption, D_c is important and the kinetics will change with temperature level. However, the necessity of including energy conservation equations will depend in addition upon the heat capacity characteristics of the system. As mentioned, the high heat capacity of liquid streams makes it much less likely that important temperature variations, for example between fluid and solid, will develop in aqueous systems. Hence, for aqueous adsorption it appears that an isothermal treatment using only the mass conservation equations will usually be satisfactory. Further analysis of space and point processes will be based upon isothermal operation.

Importance of Space Processes

A recent analysis (*4*) of kinetic data for the adsorption of hydrocarbon gases on silica gel illustrates the relative influence of the three sequential steps outlined earlier as well as the significance of axial dispersion. The data were obtained at 50°C. for a high surface area gel (832 sq. meters/gram), and are of a chromatographic type; that is, concentration—times curves were observed in the effluent gas from an adsorption bed in response to a square wave input to the bed. Rate coefficients were evaluated by comparing the chromatographic curve with the solution of Equations 1, 2, 3, 4, and 5. The importance of axial dispersion depends upon the particle diameter and velocity of the fluid stream, that is, the variables in the Peclet number. In the chromatographic analysis the effect is best viewed as a plot of the second moment of the effluent peak *vs.* the reciprocal of the square of the velocity. For certain operating conditions such graphs are straight lines. An illustration is Figure 1, in which data for the reduced second moment of propane are shown for three particle sizes. The part of the total resistance associated with axial dispersion at any velocity v is equal to the value of the ordinate at $1/v^2$ less its value at $1/v^2 = 0$. The fraction of the resistance associated with axial dispersion is this difference in ordinates divided by the ordinate at $1/v^2$. For the conditions of Figure 1, axial dispersion is observed to be a significant factor for all velocities at the smallest particle size. However, for the particles of 0.50 mm. radius at high velocities, the effect is small, but hardly negligible at the highest velocities studied.

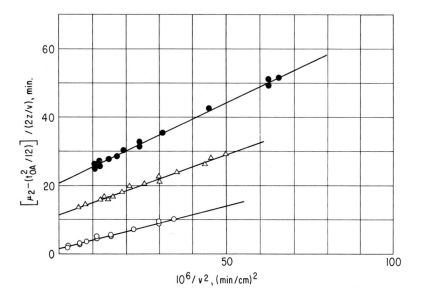

Figure 1. Chromatography of propane (50°C.) and dependence of $[\mu_2 - (t^2_{oA}/12)]/2(z/v)$ on $1/v^2$

○ $R = 0.11$ mm.
△ $R = 0.39$ mm.
● $R = 0.50$ mm.

The three sequential steps are additive so that their importance can be shown as percentage figures. Table I shows results at 50°C. for three hydrocarbons. The effect of axial dispersion has been eliminated from these data. As particle size increases the effect of increasing diffusion path length is clearly seen. External diffusion resistance also increases with particle diameter but is not a very important factor in any case. Comparison of the results for different hydrocarbons shows the effect of k_{ads}. The lower the molecular weight the slower the rate of physical adsorption. Hence, for ethane and small particle diameter the local adsorption step accounts for nearly half of the total resistance. For smaller particle sizes the trend continues until the point process can control global kinetics. This is illustrated later with some data on chemisorption, generally a slower process than physical adsorption. For aqueous solutions involving rapid adsorption or reaction at the site, for example, ion exchange processes, global kinetics are normally controlled by external and internal diffusion. This situation corresponds closely to butane adsorption on the larger particle size in Table I.

The shape of breakthrough curves provides a qualitative description of the controlling processes in an adsorption bed. Equations 1, 2, 3, 4, and 5 with proper boundary and initial conditions can be compared with measured breakthrough curves to evaluate quantitatively the resistances chosen to describe the process. This is conveniently illustrated with data (9, 10) for the adsorption of nitrogen on 96% silica glass (Vycor) at liquid nitrogen temperature. For this physical adsorption, it is expected that either external or internal diffusion would control the kinetics. Figure 2 shows observed breakthrough curves for two particle sizes and also predicted curves based upon external diffusion controlling the date. The predicted curves do not show a large enough effect of particle diameter to fit the observed data. Figure 3 shows the same type of data for four particle sizes and predicted curves based upon internal diffusion controlling the kinetics (with a fixed value for D_e). This model fits the data rather well.

Table I. Contributions to the Second Central Moment—Adsorption on Silica Gel at 50°C.

	Ethane		Propane		Butane	
	$R=$ 0.11mm	$R=$ 0.50mm	$R=$ 0.11mm	$R=$ 0.50mm	$R=$ 0.11mm	$R=$ 0.50mm
A Adsorption Resistance, %	44.5	4.0	37.7	3.0	16.0	1.0
B Intraparticle Diffusion Resistance, %	54.8	95.0	61.5	95.9	81.4	95.1
C External Diffusion Resistance, %	0.7	1.0	0.8	1.1	2.6	3.1

For chemisorption the point adsorption process is more likely to be significant. The pertinent variable to investigate in this case is temperature level. In Figure 4 breakthrough curves are illustrated for ethyl alcohol on silica gel (11). The effect of temperature is very strong. For the smallest size (0.0099 cm. radius) at 155°C. the curve is nearly vertical and corresponds to a very rapid rate. The fact that the curves for the same temperature but different sizes are not the same suggests that internal or external diffusion resistance also is important. A mathematical analysis of the curves gives evidence that a model based upon point adsorption and internal diffusion satisfactorily represents the data. Figure 5 shows that at 90°C., the breakthrough curves for the two smallest particles are essentially the same, indicating that point adsorption controls the overall kinetics. In this situation the rate is proportional to the total number of sites or to the mass of adsorbent, regardless of particle size. At 155°C. (Figure 6) the rate constant for the point process has become so much larger that the breakthrough curves vary with particle size for the whole range of particles.

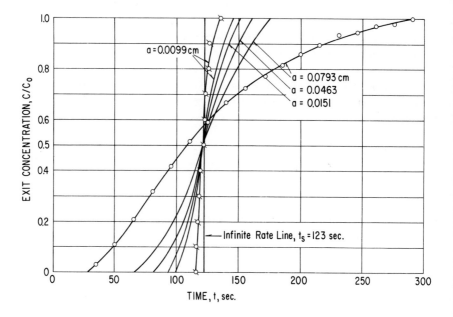

Figure 2. *Breakthrough curves for adsorption of nitrogen—external diffusion controlling*

——— *Calculated from Equation 32*
—O— *Experimental data*

In aqueous systems involving electrolytes, the rapid point process had led to equilibrium treatments (*17*), even for space steps. Considerable insight into the design of adsorption beds can be gained by treating both external and internal diffusion steps, as well as point adsorption, as occurring at equilibrium. This procedure has been particularly helpful for multicomponent systems (*2, 5, 9, 15, 18*). With respect to kinetics it has long been recognized (*1, 3, 6, 7, 16*) that diffusion steps control the kinetics of ion exchange processes. Of particular interest is the effect of the electric gradient, induced by the concentration gradient, on intraparticle mass transfer. Helfferich and Plesset (*6, 7*) have used the Nernst-Planck equation for the flux, which includes the contribution attributed to electric gradient, in conservation equations for spherical particles and for slab geometry. Assuming this internal diffusion step controls the process, the non-linear equations were integrated numerically. The results show that the electric effect depends upon the ratio of the diffusivities of the counter-moving ions D_A/D_B, and to a slight extent upon the ratio of electric charges of the two ions. Normally intraparticle diffusion has been treated in such systems using Fick's law with a single concentration-independent diffusivity for a binary system. Using the Nernst-Planck flux relationship, the rate of exchange is predicted to increase when the faster ion (for example H^+ in the H^+–Li^+ system) is in the exchange resin initially. When the ratio $D_A/D_B = 10$ the time required for 90% exchange differed by a factor of three from that based upon the usual Fick's law result.

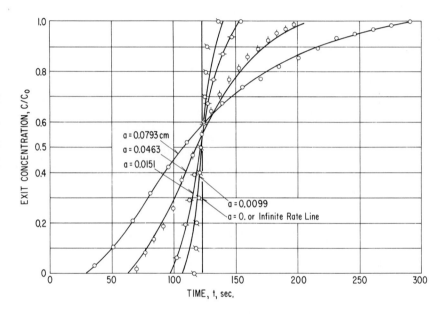

Figure 3. *Breakthrough curves for adsorption of nitrogen—internal diffusion controlling*

    -o- o o *Experimental data*
    ——— *Computed results, Equation 25*

Helfferich (4) tested the essential features of the Nernst-Planck concept by experimental measurements with the H^+–Na^+ ion pair. Diffusivities of each ion in the exchanger were evaluated from independent conductivity measurements. These diffusivities were then used to predict exchange rates and concentration profiles within the resin using the previously developed theory (6, 7). The agreement was good, and in particular the experimental data confirmed that the rate of exchange was larger when the ion of greatest mobility (H^+) was originally in the exchanger and Na^+ was in solution.

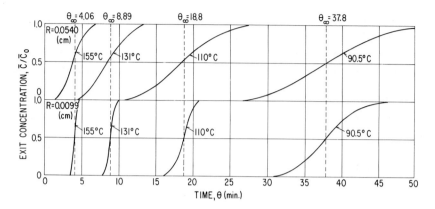

Figure 4. Breakthrough curves for adsorption of ethyl alcohol on silica gel

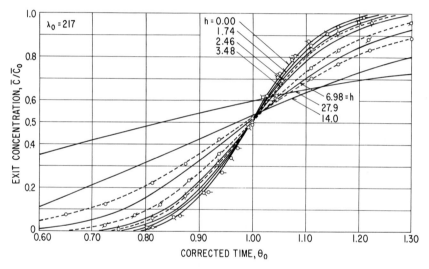

Figure 5. Converted breakthrough curves for adsorption of ethyl alcohol on silica gel at 90.5°C.

Expt. data: particle radii
o 0.0793 cm.
‚o′ 0.0540 cm.
ȯ 0.0311 cm.
ọ̀ 0.0151 cm.
-o- 0.0099 cm.

If Fick's equation is used and the diffusivities of the ion pair D_{AB} and D_{BA} are assumed to be equal, as for binary gas diffusivities, predicted exchange rates are the same, regardless of direction of diffusion. Thus, the newer theory taking into account the electric potential is clearly an improvement over the Fick's law approach. However, if an empirical view of this simple theory is used, allowing each diffusivity to assume a value determined by the data, exchange rates can be adequately represented. Hering and Bliss (8) studied the exchange rates of six ion pairs on Dowex 50W and found that the fractional conversion vs. time results could be represented with equal accuracy by either theory. However, the empirically established values for the diffusivities were an order of magnitude different for the counter diffusing ions and probably had little physical meaning.

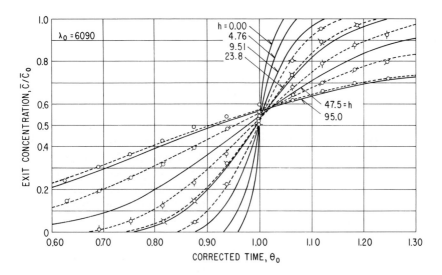

Figure 6. Converted breakthrough curves for adsorption of ethyl alcohol on silica gel at 155°C.

Expt. data: particle radii
o   0.0793 cm.
₀′ 0.0540 cm.
φ   0.0311 cm.
ₒ   0.0151 cm.
-o- 0.0099 cm.

Methods of Evaluating Space and Point Resistances

A common method of assessing the relative importance of internal diffusion and point adsorption resistances is to measure, as a function of time, the uptake of adsorbent from a solution containing solid particles. Batch data of this type taken at different temperatures and particle sizes can usually be analyzed so as to establish the importance of internal resistances. However, some types of diffusion have relatively high activation energies so that the separation is complex. Also, in such methods care must be taken to ensure rapid motion of the fluid with respect to the particles, for example by stirring, in order to eliminate external diffusion resistance. This is particularly important if the results of batch measurements are to be used in designing a fixed-bed adsorber, because external diffusion resistance is difficult to maintain the same in batch and flow systems.

Two relatively new methods have been developed which analyze directly the performance of adsorption beds. Both are based upon comparing solutions of Equations 1, 2, 3, 4, and 5 with experimental data and differ only by the initial condition applied to the bed. In the one the input is a step function of adsorbent so that breakthrough curves, such as shown in Figures 2, 3, 5, and 6 represent the concentration in the effluent from the bed. In the other the input function is a square wave, corresponding to chromatographic techniques. In this case the first and second moments can be conveniently analyzed for the rate coefficients of space and point processes. The method is now limited to linear adsorption isotherms, but frequently experimental conditions can be chosen so that this restriction is fulfilled. Adsorption rate coefficients so determined should be valid over any concentration range, whether the isotherm is linear or not, as long as the coefficients are independent of composition. The first moment, measuring the center of gravity of the effluent peak, can be related to the equilibrium constant for the point adsorption step, and the second and higher moments to the rate coefficients for axial dispersion, external and internal diffusion, and point adsorption. Normally it is not possible to evaluate moments higher than the second with enough precision to be useful. However, by making measurements for different velocities and particle sizes, under carefully chosen conditions, sufficient data are obtainable to evaluate the space and point parameters. The method has several advantages:

    1. Rate coefficients are determined in an adsorption bed, that is for the same geometry that they would be used for in design purposes.

    2. Equilibrium and rate coefficients can be established simultaneously in the same apparatus.

    3. The apparatus is simple and data are obtained rapidly.

Summary

The intent of this paper is to point out that physical or space processes, which usually influence and frequently control kinetics of adsorption in aqueous systems, can be represented effectively by quantitative models. The rate coefficients in such models are more meaningful than those associated with schemes which do not recognize space processes. Published reports have frequently analyzed data by a chemical model, but in such instances the "reaction rate" constants are found to be functions of particle size or fluid flow rate, as pointed out by Griffin and Dranoff (3).

The application of adsorption kinetics is usually the design of processes for removing or separating components from a fluid stream; a process for which the fixed-bed adsorber is well suited. Recently, analysis methods have been developed by which it is possible to evaluate the relative importance of space and point processes, and obtain numerical values of rate coefficients, from measurements on laboratory-scale fixed beds. This approach facilitates the design of large scale equipment, and in the case of the chromatographic method, possesses other advantages as a means of evaluating rate coefficients.

Finally, intraparticle diffusion appears to be an important factor in adsorption kinetics for many types of systems. In the past it has been customary to define such mass transfer quantitatively in terms of an effective diffusivity. However, even in gas–solid systems more than one process can be involved for porous particles. Thus, two-dimensional migration on the pore surface, surface diffusion, is a potential contribution. Liquid systems appear to be more complex, and, with electrolytes, it has been shown that the electric potential induced by counter-diffusing ions should be taken into account. A realistic description of intraparticle mass transfer in such cases requires more than a single rate coefficient for a binary system.

Symbols

C Specific heat; C_f for fluid; C_p for particle; cal./(gram)(°K.)
c Concentration of adsorbable gas in the interparticle space, mole/ml.
c_i Concentration of adsorbable gas in the intraparticle space, mole/ml.
c_{ads} Concentration of adsorbed gas per unit weight of adsorbent, mole/gram
D_e Effective intraparticle diffusion coefficient, cm.²/sec.
h_f External heat transfer coefficient, cal./(sec.)(cm.²)(°K.)
E_x Effective axial dispersion coefficient, based upon total cross-sectional area of bed, cm.²/sec.
ΔH Heat of adsorption, cal./mole
k_e Intraparticle effectivity thermal conductivity, cal./(cm.)(sec.)(°K.)
k_x Axial effective thermal conductivity of bed, cal./(cm.)(sec.)(°K.)
k_{ads} Adsorption rate constant, ml./(gram)(sec.)
k_f External mass transfer coefficient, cm./sec.
K Adsorption equilibrium constant, ml./gram
r Radius coordinate in spherical particle, cm.
R Radius of particle, cm.

R_v Global adsorption rate, mole/(sec.)(ml.)
R_p Point adsorption rate, mole/(sec.)(ml.)
Q_p Point heat release in particle, cal./(sec.)(ml.)
Q_v Global heat release cal./(sec.)(ml.)
t Time, sec.
T Absolute temperature in fluid stream; T_i = temperature within particle, °K.
v Linear velocity of fluid in the interparticle space, cm./sec.
z Axial coordinate in the bed, cm.
α Interparticle void fraction
β Intraparticle void fraction
ρ Density; ρ_f for fluid; ρ_p for particle, gram/ml.

Acknowledgment

The financial assistance of the American Chemical Society, PRF Grant No. 1633, is gratefully acknowledged.

Literature Cited

(1) Boyd, G. E., Adamson, A. W., Myers, L. S. Jr., *J. Am. Chem. Soc.* **69**, 2836 (1947).
(2) Glueckauf, E., *Proc. Roy. Soc. (London)* **A186**, 35 (1946).
(3) Griffin, R. P., Dranoff, J. S., *A.I.Ch.E. J.* **9**, 283 (1963).
(4) Helfferich, F. G., *J. Phys. Chem.* **66**, 39 (1962).
(5) Helfferich, F. G., *Ind. Eng. Chem. Fundamentals* **6**, 362 (1967).
(6) Helfferich, F. G., Plesset, M. G., *J. Chem. Phys.* **28**, 418 (1958).
(7) *Ibid.*, **29**, 1064 (1958).
(8) Hering, B., Bliss, H., *A.I.Ch.E. J.* **9**, 495 (1963).
(9) Klein, G., Tondeur, D., Vermeulen, T., *Ind. Eng. Chem. Fundamentals* **6**, 339 (1967).
(10) Masamune, S., Smith, J. M., *A.I.Ch.E. J.* **10**, 246 (1964).
(11) *Ibid.*, **11**, 41 (1965).
(12) Schlogl, R., Helfferich, F., *J. Chem. Phys.* **26**, 5 (1957).
(13) Schneider, P., Smith, J. M., *A.I.Ch.E. J.* **14**, 762 (1968).
(14) Smith, J. M., "Chemical Engineering Kinetics," McGraw-Hill Book Co. Inc., New York, 1956.
(15) Tondeur, D., Klein, G., *Ind. Eng. Chem. Fundamentals* **6**, 35 (1967).
(16) Vassiliou, B., Dranoff, J. S., *A.I.Ch.E. J.* **8**, 248 (1962).
(17) Vermeulen, T., "Advances in Chemical Engineering," Vol. 2, Academic Press, New York, 1958.
(18) Walter, J. E., *J. Chem. Phys.* **13**, 299 (1945).

RECEIVED October 26, 1967.

3

Two-dimensional Monomolecular Ion Exchangers

Kinetics and Equilibrium of Ion Exchange

M. DE HEAULME, Y. HENDRIKX, A. LUZZATI, and
L. TER MINASSIAN-SARAGA

Centre National de la Recherche Scientifique, Physico-Chimie des Surfaces et des Membranes, 45, rue des Saints-Peres, Paris VI°, France

The adsorption of several cationic soaps and of the $PoCl_6^{2-}$ ions at the surface of their aqueous solutions is measured. From the amount of adsorption we deduce the coefficient of selective exchange (selectivity). This coefficient depends on the chemical nature (the hydrophobic character in particular) of the organic cations which constitute the sites of the two-dimensional ion exchanger, and of the distance between these sites. Under our experimental conditions the rate of the ion exchange by the two-dimensional ion exchanger is controlled by the diffusion of $PoCl_6^{2-}$ towards the surface. This diffusion takes place through a diffusion layer located below the two-dimensional ion exchanger.

The first measurements of the adsorption of dissolved substances at the air-solution interface, or by soap films, by using molecules labelled with radioactive elements, were performed by Dixon et al. (6) and by Hutchinson (11).

Dixon et al. measured the emission into air, above an aqueous solution of the surface active compound, of soft β-rays of a ^{35}S labelled soap.

They studied the kinetics of the adsorption of the soap anions as a function of time and obtained the equilibrium values for the adsorbed layer density.

It was found that the adsorption might be controlled by the diffusion of the adsorbing species.

The adsorption of small ions (sulfate ions) at the surface of solutions of surface active agents (cationic, anionic, and non-anionic) was measured by the same authors (13, 14, 15) using the radiotracer method. Lastly, they demonstrated the displacement of one adsorbed soap by a second, more active surface agent, present in the solution (12).

The method used by Hutchinson (11) allows the measurement of the composition of soap films, the constituents of which are labelled with radioactive elements. Such a film may be formed by raising a platinum ring through the surface of the solution of the soap. The radiation emitted by the tagged molecules of the film is measured and the concentration of the last ones in the film are calculated.

The methods used by Aniansson et al. (1, 2, 3), Nilson (4, 5) and ourselves (7, 20, 21) are related to the method of Dixon et al. (6), while the method of Shinoda and all (19) is based on that of Hutchinson (11).

Most of the authors (2, 3, 7, 19, 21) who use labelled molecules are concerned with the process of ion exchange by adsorbed ionized soap molecules and calculate the separating factors or the coefficient of selectivity for the exchanging cations (2, 3, 19) or anions (7, 21).

The authors of the present paper (7, 10, 21) used the planar soap monolayers as a simple model for the complex surface of a porous ion exchanging resin. Thus, a study of the effect of the separation between the ionized sites of a "model" exchanger on its selectivity between two competing counterions can be attempted.

The suggested model is consituted by the aqueous solution of cationic soaps, containing HCl ($2M$) and traces (10^{-7}–$10^{-8}M$) of a ^{210}Po salt which, under our conditions, is the compound $^{210}PoCl_6H_2$ (4, 5).

At the surface of the solution the cationic soap constitutes a positive layer of organic cations neutralized by Cl^- and $PoCl_6^{2-}$ ions; ^{210}Po emits α–rays. The surface density as well as the kinetics of the adsorption of $PoCl_6^{2-}$ are determined by measuring the radioactivity above the surface of the solution. An analogous technique is used to determine the density of the soap adsorbed at the air-solution interface, when the soap is labelled with ^{14}C. Therefore, these measurements allow a direct analysis of the composition of the adsorbed soap films at equilibrium and during their formation.

From the equilibrium values we calculated the coefficient of selectivity (9) and the corresponding free energy and determind their variation as a function of the separation between the ionized sites.

A preliminary report (10) and a more detailed one (7) of the techniques employed are published elsewhere. Four cationic soaps are used as follows: dodecyltrimethylammonium bromide (C_{12}TABr,I); hexadecyltrimethylammonium bromide H(C_{16}TABr,II)*, hexadecylpyridinium chloride (C_{16}PyCl,III)*, and hexadecyldimethylethylolammonium bromide (C_{16}CholBr,IV). The last compound is related to choline bromide, one methyl of which is replaced by a hexadecyl radical. The soaps marked with an asterisk are labelled ^{14}C.

Results and Discussion

Kinetics of Adsorption (8) (Results). (A) THE IONS $PoCl_6^{2-}$. The curves obtained by plotting the intensity of the α-rays emitted by ^{210}Po adsorbed by the soap film vs. time conform to the law represented by the Figure 1. The slope k of the line shown on Figure 1 is the constant of the rate of exchange of these ions by the pre-existing film of the neutral adsorbed soap. No such adsorption can be found for an anionic soap or oleic acid monolayers.

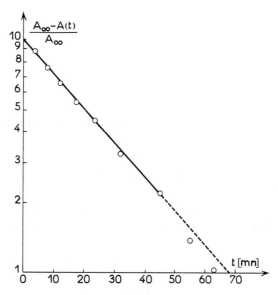

Figure 1. Kinetics of the adsorption of $PoCl_6^{2-}$ ions

$A_\infty = \alpha$-ray emission at equilibrium. $A(t)$: instant α emission. t: time. $k = $ slope of the line. $c_{C_{16}TABr} = 2 \times 10^{-5}M$; HCl 2M

(B) THE SOAP FILMS. The formation of the soaps films II and III is followed by measuring the variation of the intensity of the β-rays emitted by the ^{14}C atoms labelling the organic cations of the soaps. Their equilibrium surface density is attained long before that of the $PoCl_6^{2-}$ ions.

Equilibrium of Adsorption. (A) THE SOAP FILMS. The isotherms of adsorption of the organic cations of the soaps are reproduced on the Figure 2. Classical methods (20) are used to obtain the surface densities of the soaps I and IV.

(B) THE $PoCl_6^{2-}$ IONS. Considering that the maximum number of positive sites is given by the surface density δ of the organic cations, the amount of adsorption of the ions $PoCl_6^{2-}$, $\delta_{PoCl_6^{2-}}$, does not exceed 1% of this maximum value. Under these conditions $\delta_{PoCl_6^{2-}}$ is proportional to the concentration in bulk of the same ions $c_{PoCl_6^{2-}}$ therefore the distribution coefficient of this ion

$$K = \frac{\delta_{PoCl_6^{2-}}}{c_{PoCl_6^{2-}}} \qquad (1)$$

is independent of $c_{PoCl_6^{2-}}$. However, it varies with δ (7, 10).

The coefficient of selectivity (9) α corresponding to the following reaction of exchange

$$\overline{2\ Cl^-} + PoCl_6^{2-} \rightleftharpoons \overline{PoCl_6^{2-}} + 2\ Cl^- \qquad (2)$$

can be determined from our data. If one assumes that $\delta_{Cl^-} = (\delta - 2\delta_{PoCl_6^{2-}})$ and, as $\delta_{PoCl_6^{2-}} \ll \delta_{Cl^-}$ and $c_{PoCl_6^{2-}} \ll c_{Cl^-}$, it can be shown that α is equal to:

$$\alpha \approxeq \frac{\delta_{PoCl_6^{2-}}}{c_{PoCl_6^{2-}}} \times \frac{c_{Cl^-}}{\delta} = K\frac{c_{Cl^-}}{\delta} \qquad (3)$$

From Equation 3 we calculate the free enthalpy of exchange $\Delta G_{2Cl}^{PoCl_6^{2-}} = -RT\ \ln\alpha$.

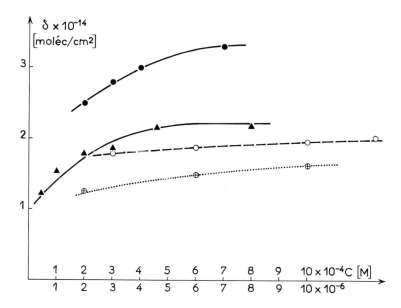

Figure 2. Surface density of δ of the adsorbed soap films vs. its concentration c in the solution

HCl 2M; 25°C. △ ($C_{12}TABr$); ⊙ ($C_{16}CholBr$); ○ ($C_{16}TABr$); + ($C_{16}PyCl$); $C_{12}TABr$ (0–10^{-3}M); other soaps (0–10^{-5}M)

Figure 3 shows a plot of this vs. the quantity $(1/\sqrt{\delta})$ which is proportional to the separation between the positive charges of the adsorbed soap monolayer.

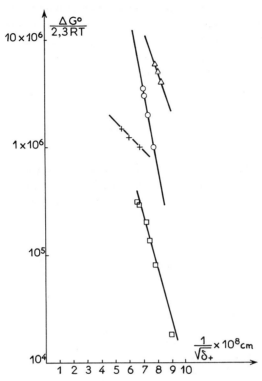

Figure 3. The free enthalpy of ion exchange $\Delta G/RT$ vs. the separation $(1/\sqrt{\delta})$ (Å) between the positively charged sites 25°C.

From bottom to top, the soaps $C_{12}TABr$, $C_{16}CholBr$, $C_{16}TABr$, $C_{16}PyCl$

For a given separation between the charged sites and for the ions $PoCl_6^{2-}/Cl^-$ the chemical nature of the adsorbing sites may affect the ionic selectivity strongly. The hydrophobic character of the substituted quaternary ammonium seems to favor the selection of the $PoCl_6^{2-}$ ion (compare the soaps I and III). If we consider the exchange $PoCl_6^{2-}/Cl^-$ as an "in situ indicator" of the hydrophobic character of the adsorbing site our results (Figure 3) allow us to class the polar groups of II, III, and IV according to the intensity of this character, as follows: pyridinium > trimethylammonium > dimethylethylolammonium.

Kinetics of Adsorption. DISCUSSION. The interpretation of the results reproduced on the Figure 1 is attempted on the same lines as the desorption of slightly soluble monolayers (20).

It is assumed that during the exchange, the "adsorbed" ions $PoCl_6^{2-}$ distribute locally between the soap monolayer and a very thin region of neighboring liquid substrate. The local distribution ratio is equal to K of the Equation 1 and varies with time following the variation of $\delta_{PoCl_6^{2-}}$. In this thin region the concentration $c_{PoCl_6^{2-}}$ is lower than in the bulk of the substrate and $PoCl_6^{2-}$ ions migrate across a thick diffusion layer ϵ until the concentration of the ions $PoCl_6^{2-}$ becomes uniform throughout all the liquid substrate.

Under these conditions the constant k of the rate of adsorption is equal to:

$$k = \frac{1}{K} \frac{D}{\epsilon} \qquad (4)$$

where $D = 0.78$ cm.^{2}/day (*17*). The Equation 4 is verified by the soaps II, III, and IV (*see* Figure 4). The hydrodynamic constant (D/ϵ) is independent of the natuer of the "adsorbent" (soap film).

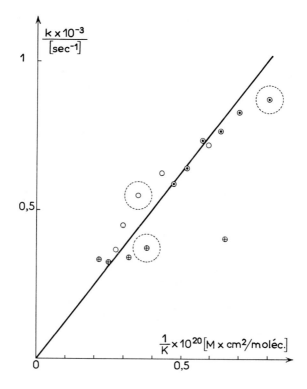

Figure 4. The constant of the adsorption rate vs. (1/K)

K = *distribution coefficient of $PoCl_6^{2-}$ (Equation 1)* ⊙ $C_{16}CholBr$; ○ $C_{12}TABr$; ⊕ $C_{16}PyCl$

From it a value of $\epsilon \cong 0.4$ mm. is calculated using the value of D (17).

Conclusion

The kinetics and the equilibrium of adsorption of anions $PoCl_6^{2-}$ by ionized monolayers of cationic soaps have been studied and both phenomena are analogous to those observed with the polymer ion exchangers (9).

It is suggested that the soap films may be considered as monomolecular model ion-exchangers (two-dimensional) and made use of for the study of particular effects contributing to the usual complex process of ion exchange. As a first application, the authors study the effects of the separation between the exchanging sites and of the degree of their hydrophobic character on the selectivity of ion exchange.

Literature Cited

(1) Aniansson, G., *J. Phys. Coll. Chem.* **55**, 1286 (1951).
(2) Aniansson, G., Steiger, N. H., *J. Chem. Phys.* **21**, 1299 (1953).
(3) Aniansson, G., Steiger, N. H., *J. Phys. Chem.* **58**, 228 (1954).
(4) Bagnall, K. W., d'Eye, R. W., Freeman, J. W., *J. Chem. Soc.* **1955**, 2320.
(5) *Ibid.*, **1956**, 2770.
(6) Dixon, J. R., Weith, A. J., Argyle, A. Jr., Salley, D. J., *Nature* **163**, 845 (1949).
(7) de Heaulme, M., Hendrikx, Y., Luzzati, A., Ter Minassian-Saraga, L., *J. Chim. Phys.* **64**, 1363 (1967).
(8) de Heaulme, M. (to be published).
(9) Helferich, E., "Ion Exchange," McGraw-Hill, New York, 1962.
(10) Hendrikx, Y., Luzzati, A., Ter Minassian-Saraga, L., *J. Chim. Phys.* **59**, 481 (1962).
(11) Hutchinson, E., *J. Coll. Sci.* **4**, 600 (1949).
(12) Judson, G. M., Argyle, A. A., Salley, D. J., Dixon, J. K., *J. Chem. Phys.* **18**, 1302 (1950).
(13) *Ibid.*, **19**, 378 (1951).
(14) Judson, G. M., Lerew, A. A., Dixon, J. K., Salley, D. J., *J. Chem. Phys.* **20**, 519 (1952).
(15) Judson, G. W., Lerew, A. A., Dixon, J. K., Salley, D. J., *J. Phys. Chem.* **57**, 916 (1953).
(16) Nilson, G., *J. Phys. Chem.* **61**, 1135 (1957).
(17) Servigne, M., *J. Chim. Phys.* p. 31 (1934).
(18) Shinoda, K., Nakanishi, J., *J. Phys. Chem.* **67**, 2547 (1943).
(19) Shinoda, K., Ito, K., *J. Phys. Chem.* **65**, 1499 (1961).
(20) Ter Minassian-Saraga, L., *J. Chim. Phys.* **53**, 355 (1956).
(21) Ter Minassian-Saraga, L., Luzzati, A., *J. Chim. Phys.* **59**, 481 (1962).

RECEIVED November 24, 1967.

4

Chromatographic Behavior of Interfering Solutes

Conceptual Basis and Outline of a General Theory

F. HELFFERICH

Shell Development Company, Emeryville, Calif. 95608

The basis of a theory for interfering solutes, applicable to systems with any number of species and arbitrary initial and influent conditions, is outlined. The key concept is that of "coherence": a composition profile is coherent if, at a given time, all concentrations coexisting at any location have the same velocity. Given sufficient time and distance for undisturbed development, coherence is attained from any arbitrary initial conditions. Coherent composition sequences, mapped in a coordinate system with concentrations as coordinates, are called "composition paths." A simple example illustrates the use of composition paths in predicting development behavior. For systems with constant separation factors, the composition-path grid can be orthogonalized by the so-called h-transformation, which greatly simplifies the mathematical treatment.

The dynamic behavior of a species passing through a chromatographic column is chiefly determined by the distribution of the species between the stationary phase—e.g., sorbent, ion exchanger—and the mobile phase (moving gas or liquid). In conventional theories of chromatography it is usually assumed that the various species in the column do not affect one another's distribution. Mathematically, in equilibrium, the stationary-phase concentration y_i of a species i then is a function only of the mobile-phase concentration x_i of the species:

$$y_i = f(x_i) \qquad (1)$$

This assumption conveniently permits multicomponent cases to be treated as composites of single-component cases: the behavior of each species can be calculated separately and the results be superimposed on one another. The assumption is quite acceptable for analytical chromatography at low concentrations and low degree of sorbent loading; but becomes untenable at high concentrations or in ion exchange with high conversion, because the solute species then affect one another's sorption behavior as they compete for the limited number of available sorption "sites." In equilibrium, the stationary-phase concentration of species i then depends on the mobile-phase concentrations of all species present rather than only on that of i:

$$y_i = f(x_1, x_2, \ldots x_n) \qquad (2)$$

The term "interfering solutes" has been coined to characterize this situation. The seemingly minor complication has serious consequences for the theoretical treatment and may be said to add a new dimension to chromatographic behavior. An examination of the effects caused by solute interference is the subject of the present communication.

Scope and Premises

A brief introduction to a new theory developed for interfering solutes will be given. The basic and new concepts will be outlined and illustrated. Only a few highlights can be given, and space does not permit to include proofs. For a comprehensive coverage, details, proofs, and a complete mathematical treatment the reader is referred to a forthcoming monograph (*11*).

To bring out clearly the effects arising exclusively from solute interference, all disturbances well known to occur in chromatographic columns —deviations from local equilibrium, finite mass-transfer rates, axial diffusion, eddy dispersion, flow maldistribution, fingering, pressure drop, deviations from isothermal behavior—will be disregarded. The effects of such disturbances are essentially the same in systems with interference as in any others and can be accounted for by later corrections. For simplicity, only what may be called "competitive" sorption will be examined: addition of a sorbable species at constant concentrations of the others, or replacement of a species by another with higher affinity for the sorbent, decreases the distribution ratios y_i/x_i of all species because of the increased competition for the sorption sites. Furthermore, the exchange of species between the stationary and mobile phases will be assumed to be stoichiometric, as in ion exchange with dilute solutions. While the theory also covers "synergistic" (as opposed to "competitive") and nonstoichiometric sorption, the ensuing complications would tend to obscure the main features to be discussed here. In all other respects, the treatment is general and, in particular, covers systems with any number of species and with any arbitrary uniform or nonuniform initial concentrations profiles and constant or varying influent compositions.

The following conventions as to nomenclature will be adopted. The exchanging sorbable species are numbered 1,2, . . . ,n in the sequence of decreasing affinity for the stationary phase. The number of species is n. Stationary- and mobile-phase concentrations y_i and x_i are so normalized that

$$\sum_{i=1}^{n} x_i = 1 \quad \text{and} \quad \sum_{i=1}^{n} y_i = 1 \tag{3}$$

Binary separation factors are defined in the usual way as

$$\alpha_{jk} \equiv \frac{y_j x_k}{x_j y_k} \tag{4}$$

Conceptual Basis: Species Velocities and Concentration Velocities

The conceptual basis of the theory is provided by an examination of the velocities of species and concentrations in a chromatographic column. A distinction is made between "species velocities" and "concentration velocities." The species velocity is defined as the (local average) rate of advance of the molecules or ions of the respective species, and would be observed in an experiment in which the progress of, say, a radioactive tracer is measured. The concentration velocity, in contrast, is defined as the rate of advance of a given concentration of the respective species—*i.e.*, the concentration velocity of a given concentration x_i of species i is defined as $(\partial z/\partial t)_{x_i}$ (where z = distance from column inlet, and t = time). The species velocity and the concentration velocity, for the same species, usually differ from one another as they refer to physically different phenomena. They are formally analogous to the particle velocity and wave velocity in nonlinear acoustics (*13*). A cruder, but more readily perceived analogy is with a flood wave on a river; the flood wave reaches the estuary far in advance of the rain-water molecules which had caused the flood far upstream. An analogy from everyday life is a traffic jam as shown in Figure 1; defined as the locus of stopped cars or of highest car density, the traffic jam even moves against the traffic flow as arriving cars are forced to stop behind others while these get under way again (*10*).

In chromatography, the species velocity is determined by the ratio y_i/x_i of the absolute concentrations of the species in the two phases, as

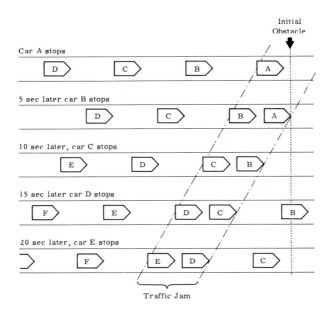

Figure 1. Traffic-jam situation on single-lane highway
(10)

only those molecules advance which are in the mobile phase. In contrast, the concentration velocity is determined by the ratio of the concentration variations of the species in the two phase—*i.e.*, by the derivative dy_i/dx_i if there is no interference. The basic difficulty arising in systems with interference is that a derivative dy_i/dx_i does not exist because y_i is no longer a function of x_i alone. Rather, the ratio of the concentration variations, and thus the concentration velocity of a species, can in principle assume any positive or even negative value, depending on the concomitant concentration variations of the accompanying species.

The Key Concept: Coherence

The key concept of the theory is that of "coherence." If one allows completely arbitrary concentration profiles of the species, one will usually find that the species concentrations coexisting at any given time and location in the column have different velocities. Compositions then exist only momentarily, as their constituent concentrations advance at different rates and thus part company. Composition profiles with this property will be called "noncoherent." In contrast, a profile will be called "coherent" if, at a given time, all concentrations coexisting at any location have the same velocity, so that compositions remain conserved as they migrate.

The mathematical condition for coherence is that, at the time under consideration, the requirement

$$(\partial z/\partial t)_{x_1} = (\partial z/\partial t)_{x_2} = \ldots = (\partial z/\partial t)_{x_n} \tag{5}$$

is met at any location z. Coherent profiles may, and usually do, contract or expand, since the velocity at a given time may vary with location; sharpening and spreading of chromatographic boundaries thus takes place in coherent systems with interference in much the same way as in systems without interference. Coherence merely requires, by definition, that the concentrations constituting any composition in the column move at the same rate and thus remain together.

At first glance, coherence may appear to be a highly exceptional behavior which one could expect to find only under special and very simple operating conditions. This is not so. The concept owes its value to the fact that any arbitrary initial composition profile eventually attains coherence if given sufficient time and distance for undisturbed development—*i.e.*, development through a uniformly presaturated bed section by an influent of constant composition. This is not entirely surprising, since coherence may be viewed as a "stable" state, and noncoherence as an "unstable" one. Once a profile has become coherent, it consists, by definition, of compositions which are conserved, and thus does not by itself become noncoherent again. In contrast, in a noncoherent profile there is a continuous shift of concentrations relative to one another. That the concentrations indeed shift in such a manner as to result eventually in coherence is, however, more difficult to prove. We shall return to this question at a later stage.

From a historical point of view, it is interesting to note that previous theories have not at all questioned as restrictive a condition as coherence. Rather, for the simple boundary conditions to which they have confined themselves, they have taken an entirely coherent behavior for granted (2, 3, 4, 5, 7, 8, 9, 12, 14, 15, 16, 17, 18, 19, 20). Apparently, only Baylé and Klinkenberg (1) recognized that an unproved postulate is involved. The present theory encompasses noncoherent as well as coherent behavior and is free of any such postulate.

Properties of Coherent Systems: Affinity Cuts, Composition Paths, and Composition Velocities

A treatment of coherent systems is inherently much simpler than that of the general case and is useful in that it allows at least to predict which state a composition variation will eventually attain if developed without being disturbed.

Coherence imposes stringent restrictions on the behavior of the species concentrations. Of the long list of required properties of coherent composition profiles only one which is rather striking will be mentioned, namely, the existence of "affinity cuts." In a coherent composition profile, each location z at which concentration gradients exist has a so-called affinity cut. This cut divides the species, written in the sequence of decreasing affinity for the stationary phase, into two groups:

$$1, 2, \ldots, j \mid j + 1, \ldots, n \tag{6}$$

where the vertical line marks the cut. The significance of the cut is that the species of the two groups change their concentrations in opposite directions. Thus, the concentrations of species 1 through j may increase, and those of species $j + 1$ through n then decrease, in the direction of flow, or *vice versa*. A profile not obeying this restriction, say, one in which at some location the concentrations of species 1 and 4 increase while those of species 2, 3, 5, and 6 decrease, cannot be coherent. The rule applies to stationary-phase as well as to mobile-phase concentrations. In an n-component system with stoichiometric exchange, there are $n - 1$ possible positions of the affinity cut namely, $1 \mid 2$, $2 \mid 3$, ..., $n - 1 \mid n$. (The notation $j \mid k$ is adopted to denote a cut between species j and k.) As a rule, a coherent composition profile involves more than one affinity cut; the profile is usually composed of various separate portions having different affinity cuts, interspersed between zones of uniform composition.

Composition sequences compatible with coherence can be mapped in a coordinate system with concentrations as coordinates. The lines thus obtained will be termed "composition paths." A composition-path grid for a three-component system is shown as an example in Figure 2. (The triangular simplex is used to accommodate compositions of three species in a two-dimensional diagram; this is entirely a matter of graphical representation and has no bearing on the mathematical treatment.) There are two sets of paths, corresponding to the two possible affinity cuts $1 \mid 2$ and $2 \mid 3$. Of the infinite multitude of paths of each set, only a few at regular intervals are shown. Each composition in the diagram is at an intersection of two paths, one from each set. A topologically important feature is that the $1 \mid 2$ and $2 \mid 3$ paths originate from different portions of the border $x_2 = 0$ of the diagram; the point W separating these two portions will be called the "watershed point." The paths are linear if the separation factors are constant—*i.e.*, independent of composition. Their exact positions depend on the values of the separation factors, but their topology does not.

A model depicting composition paths in the simplex tetrahedron of a four-component system is shown in Figure 3. For systems with more than four components, one runs out of dimensions for graphical or spatial

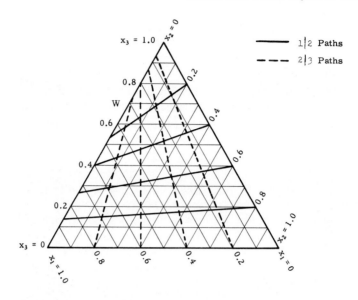

Figure 2. Composition-path grid in mobile-phase composition simplex of three-component system with separation factors $\alpha_{12} = 2$ and $\alpha_{13} = 4$

representation; nevertheless, from the $(n - 1)$-dimensional topology of the path grid the qualitative behavior for coherent cases can still be predicted with virtually no recourse to mathematics.

For coherent profiles, in which compositions remain conserved as they migrate, a "composition velocity" can be defined as the rate of advance of a given composition. The composition velocity equals the concentration velocities, which must equal one another for the profile to be coherent. The composition velocity depends not only on the composition, but also on the affinity cut. In an n-component system, in which each composition is at an intersection of $n - 1$ composition paths with different cuts, any composition has $n - 1$ potential velocities, one for each path or cut. Mathematically, to find the $n - 1$ potential velocities of a given composition is an eigenvalue problem. For any given composition, the lowest eigenvalue is for the $1 \mid 2$ cut, the next higher eigenvalue is for the $2 \mid 3$, etc., and the highest is for the $n - 1 \mid n$ cut.

Development of Noncoherent Profiles

A simple example may serve to illustrate the development of a noncoherent profile, the use of the composition-path concept in predicting the behavior even of noncoherent systems, and the physical factors responsible for eventual attainment of coherence from arbitrary initial conditions.

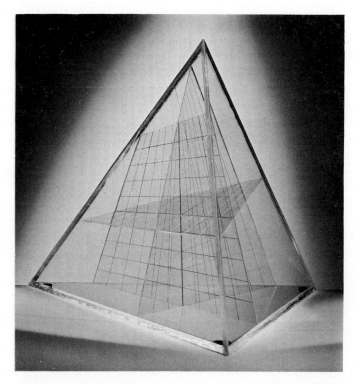

Figure 3. Model showing sets of composition paths in mobile-phase composition simplex of four-component system with separation factors $\alpha_{12} = 2$, $\alpha_{13} = 4$, and $\alpha_{14} = 8$

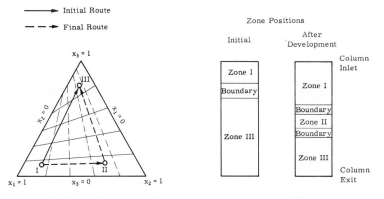

Figure 4. Development of a diffuse noncoherent boundary between zones of uniform compositions. Left: initial and final composition routes; right: positions of zones in column

Suppose the initial composition profile in the column is as shown in Figure 4. A diffuse "boundary"—*i.e.*, a zone involving concentration gradients—with appreciable concentration variations of species 1 and 3, but with almost uniform concentration of species 2, separates two zones of uniform compositions I and III. The "composition route" corresponding to this profile is shown as a solid arrow (pointing in the direction of flow) in the path grid. The noncoherent initial boundary is now developed with an eluent of composition I, so that no additional composition variations occur at the column inlet. According to the theory, the resulting undisturbed development deforms the composition route between its fixed end points I and III in such a manner that it eventually runs entirely along composition paths in numerical sequence of affinity cuts. The respective final route, after attainment of coherence, is represented by the dashed arrows. At point II in the diagram, the final coherent route changes from the $1\,|\,2$ path to the $2\,|\,3$ path; since the composition velocity of a composition is higher for the latter than for the former path, a zone of uniform composition II will grow within the profile after coherence has been attained. Thus, upon development, the initial noncoherent boundary breaks up into two coherent ones with $1\,|\,2$ cut (upstream) and $2\,|\,3$ cut (downstream), separated by a new zone of uniform composition.

It may appear surprising that development should so drastically alter the concentration profiles, in particular that of species 2, namely, generate a new zone containing this species in high concentration although its concentration in the initial profile was uniformly low. The explanation of this phenomenon provides an illustration of the physical factors which produce coherence. As Figure 4 shows, the composition I on the upstream side of the initial boundary is rich in the high-affinity species 1 and lean in the low-affinity species 3, whereas the opposite is true for the composition III on the downstream side. Accordingly, the competition for sorption sites is stronger at I than at III, with the result that the distribution ratios y_i/x_i are lower, and the species velocities therefore are higher, for all species at I than at III. For species 2, with its almost level initial profile, the higher species velocity on the upstream side results in an accumulation between the zones I and III and thus shifts the composition route in the direction corresponding to attainment of coherence. With material-balance arguments of this type one can indeed prove attainment of coherence.

The degree of detail with which the theory predicts column responses may be illustrated with a moderately complex case. Suppose that a brief pulse consisting exclusively of a single arbitrary species j is injected into an otherwise constant influent consisting of n species 1, 2, . . . , j, . . . , n. For this case the theory predicts: (1) the single influent pulse is resolved into $n - 1$ response pulses traveling at different rates; (2) the injected species j has a concentration maximum in all response pulses; (3) the concentration of the other species have maxima or minima in the response pulses as shown in Table I; (4) all pulses are skewed: the first through $(j - 1)$'th (counted in the direction of flow) have diffuse front flanks and relatively sharp rear flanks, and the j'th through $(n-1)$'th have the opposite skewness. For a ten-component system, no less than one hundred observable features of the response pattern in such a case are predicted.

Table I. Response-pulse Pattern Produced by Injection of a Single-Component Pulse of a Species j into an Otherwise Constant Influent of All Species Including j[a]

Species	1	2	3	4	...j−1	j	j+1	...n−2	n−1	n
1st pulse	−	+	+	+	+	+	+	+	+	+
2nd pulse	−	−	+	+	+	+	+	+	+	+
3rd pulse	−	−	−	+	+	+	+	+	+	+
...										
(j − 1)th pulse	−	−	−	−	−	+	+	+	+	+
jth pulse	+	+	+	+	+	+	−	−	−	−
...										
(n − 2)th pulse	+	+	+		+	+	+	+	−	−
(n − 1)th pulse	+	+	+		+	+	+	+	+	−

[a] Pulses are numbered sequentially in the direction of flow; plus and minus signs indicate concentration maxima and minima, respectively, of the respective species.

H-Function and h-Transformation

For systems with constant—i.e., composition-independent—separation factors, a much sharper tool for handling noncoherent as well as coherent cases is provided by what will be called the "*h*-transformation," which will now be outlined.

The peculiar topology of the composition-path grid in Figure 2 lets it appear possible to deform the simplex triangle, by pulling out the watershed point, into a rectangle with all paths at right angles (*see* Figure 5). The transformation is nonlinear and involves a singularity at the watershed point, but is achieved without much difficulty, even for systems with more than three components. In the general case, the $(n - 1)$-dimensional composition simplex of an n-component system is transformed into a rectangular $(n - 1)$-dimensional parallelepiped with all paths at right angles. Any composition can now be expressed in terms of its $n - 1$ coordinates in the transformed space instead of its $n - 1$ independent concentrations x_i (or y_i). For a given composition $\{x_1, x_2, \ldots, x_n\}$, the coordinate values $h_1, h_2, \ldots, h_{n-1}$ in the transformed space are obtained as the $n-1$ roots in h (all real and positive) of the equation

$$H(h, x_1, x_2, \ldots, x_{n-1}) = 0 \qquad (7)$$

where the so-called H-function (hyperplane function) is defined as

$$H(h, x_1, x_2, \ldots, x_{n-1}) \equiv \sum_{i=1}^{n-1} \left(\frac{\alpha_{1n} - \alpha_{1i}}{h - \alpha_{1i}} x_i \right) - 1 \qquad (8)$$

Conversely, the concentrations x_i constituting a composition characterized by its set of H-function roots can be calculated from the latter by

$$x_j = \prod_{i=1}^{n-1} (h_i - \alpha_{1j}) \bigg/ \prod_{\substack{i=1 \\ i \neq j}}^{n} (\alpha_{1i} - \alpha_{1j}) \qquad \text{for all } j = 1, \ldots, n \qquad (9)$$

The h-transformation is designed to orthogonalize the composition-path grid. Thus, since any path is normal to all coordinates but one, only one H-function root can vary along any path. The coherence requirement can therefore be simply expressed as

$$dh_i = 0 \qquad \text{for all } i \neq k \qquad (10)$$

where k can have any value $1, \ldots, n-1$. Along a path with $k \mid k+1$ affinity cut, the variable root is h_k. Not surprisingly, the relations for all quantities of interest, such as distribution ratios, species velocities, composition velocities, assume very simple forms when expressed in terms of H-function roots.

The basic set of differential material-balance equations for the various species in the column can also be written in terms of the h_i instead of x_i and y_i. This new set of differential equations reveals a particular property of the H-function roots: no root values other than those appearing in the initial and boundary conditions can arise anywhere in the column. The column behavior thus is completely described by the trajectories of the initial and boundary values of the roots. (The same is not true for concentrations, as the behavior of species 2 in the example in Figure 4 has shown.)

A "root velocity" can be defined as the rate of advance of a given value of an H-function root. For any given composition, roots with lower index numbers have lower velocities. An arbitrary initial noncoherent boundary involving variations of all roots thus is resolved, upon undisturbed development, into separate variations of the roots. This is shown by schematic trajectories of root values in a distance-time diagram in Figure 6. After resolution, each trajectory bundle involves variation of

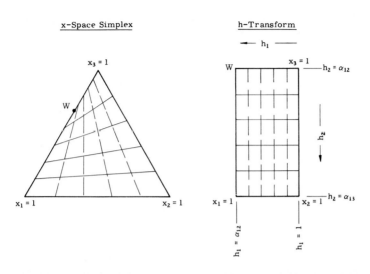

Figure 5. Three-component composition simplex with path grid, and its h-transform. (For $\alpha_{12} = 2$ and $\alpha_{13} = 4$)

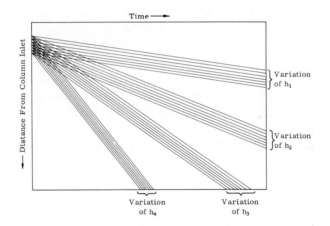

Figure 6. Root trajectories for development of noncoherent boundary in five-component system, illustrating resolution into separate variations of individual roots (schematic)

only one root and thus corresponds to a coherent profile portion. (For clarity, the trajectories in the schematic diagram have been drawn parallel and linear; this is correct only if the initial boundary involves but a very small composition change.) The resolution of root variations from one another provides a proof for attainment of coherence from arbitrary initial conditions. One may say that, in a system with interference, variations of

H-function roots are resolved from one another in the same manner as species are in conventional elution development.

Outlook

Although the theory is based entirely on the established differential material-balance equation of chromatography and involves no additional postulates, a variety of new results are obtained. The treatment is generally applicable to any initial and boundary conditions and thus is not restricted to actual chromatographic operations; it can be used to describe any other variants of the problem of a multicomponent fluid moving through a sorbent layer—*e.g.*, leaching of salts or fertilizer by rain water in soils. The treatment is crude, and much work remains to be done to incorporate corrections for various disturbances; such refinements, however, should await confirmation of the principal features. The known modes of chromatographic operation—elution development, displacement development, frontal analysis, and vacancy chromatography—are contained as special cases, which, however, constitute but a small fraction of the predictions to which the theory leads. Beyond these known cases, which are correctly described, the treatment must be considered as a purely theoretical construction unbiased by any experimental evidence. Literally hundreds of experiments suggest themselves as crucial tests of vital features, and the author gladly offers his assistance in any such endeavors.

On a broader scope, the theory has interesting analogies with non-linear acoustics, including shock waves. The fact that a single input signal—*i.e.*, a single composition change at the inlet—produces a multiplicity of response signals moving at different speeds is related to information theory. The mathematical formalism, including the h-transformation, is directly applicable to the physically different problem of multicomponent electrophoresis; in fact, the first use of a function of the type of the H-function apparently was made by Dole (6), in 1945, for electrophoresis. There are indications that a variety of other phenomena may be approached in a similar manner. A particular challenge is held out by the prospect that the treatment could conceivably be generalized for chemically reacting systems.

Literature Cited

(1) Baylé, G. G., Klinkenberg, A., *Rec. Trav. Chim.* **73**, 1037 (1954).
(2) Claesson, S., *Arkiv Kemi Mineral. Geol.* A **23**, No. 1 (1948).
(3) Claesson, S., *Ann. N.Y. Acad. Sci.* **49**, 183 (1948).
(4) Cooney, D. O., Lightfoot, E. N., *Ind. Eng. Chem., Process Design Develop.* **5**, 25 (1966).
(5) DeVault, D., *J. Am. Chem. Soc.* **65**, 532 (1943).

(6) Dole, V. P., *J. Am. Chem. Soc.* **67**, 1119 (1945).
(7) Dranoff, J., Lapidus, L., *Ind. Eng. Chem.* **50**, 1648 (1958).
(8) Glueckauf, E., *Proc. Roy. Soc. (London)* **A 186**, 35 (1946).
(9) Glueckauf, E., *Discussions Faraday Soc.* **7**, 12 (1949).
(10) Helfferich, F., *J. Chem. Educ.* **41**, 410 (1964).
(11) Helfferich, F., Klein, G., "Chromatographic Behavior of Interfering Soutes," Marcel Dekker, Inc., New York, in preparation.
(12) Klein, G., Tondeur, D., Vermeulen, T., *Ind. Eng. Chem., Fundamentals* **6**, 339 (1967).
(13) Landau, L. D., Lifshitz, E. M., "Fluid Mechanics," p. 367, Pergamon Press, London, 1959.
(14) Mangelsdorf, P. C., *Anal. Chem.* **38**, 1540 (1966).
(15) Offord, A. C., Weiss, J., *Discussions Faraday Soc.* **7**, 26 (1949).
(16) Sillén, L. G., *Arkiv Kemi* **2**, 477 (1950).
(17) Stalkup, F. I., Deans, H. A., *A.I.Ch.E. J.* **9**, 106 (1963).
(18) Tondeur, D., Klein, G., *Ind. Eng. Chem., Fundamentals* **6**, 351 (1967).
(19) Tudge, A. P., *Can. J. Phys.* **39**, 1600 (1961).
(20) Walter, J. E., *J. Chem. Phys.* **13**, 229 (1945).

RECEIVED October 26, 1967.

5

Adsorption of Hydrolyzed Hafnium Ions on Glass

LYNDEN J. STRYKER and EGON MATIJEVIĆ

Department of Chemistry and Institute of Colloid and Surface Science, Clarkson College of Technology, Potsdam, N. Y.

The adsorption of hafnium species on glass was found to increase with the solution pH and hafnium concentration. The effects on the adsorption of the solution preparation and age were studied and the equilibration time for the adsorption process was determined. The surface area of the glass sample was determined by the B.E.T. method using water vapor. The results are discussed in terms of the hydrolyzed hafnium(IV) species. At equilibrium, nearly monolayer coverage was obtained at pH > 4.5. Under these conditions hafnium is in the solution in its entirety in the form of neutral, soluble $Hf(OH)_4$ species. In the close packed adsorption layer the cross-sectional area of this species is 24 $A.^2$ which is nearly the same as for water on silica surfaces.

Numerous experiments dealing with the adsorption of metal ions on glass from aqueous solutions have been reported in the literature. These studies dealt, for example, with sodium (6, 11, 18), potassium (6), cesium (11, 18), thallium(I) and thallium(III) (32), silver (11), calcium (6), zirconium (29, 34, 36), ruthenium (4, 31), cerium (4), promethium (4), gold (26), bismuth (17), lead (17), polonium (17, 28, 33), thorium (24), radium (22), protactinium (35), uranium (30), and plutonium (7, 8, 9, 10, 13, 14). In the majority of the experiments radioactive isotopes were used in tracer quantities. It was established without exception that the pH was the most important parameter affecting the adsorption. In the case of simple non-hydrolyzable ions the pH effects were interpreted in terms of the hydrolysis of the glass surface. It is also generally suggested that the mechanism of adsorption of non-hydrolyzed ions is essentially an ion exchange of the adsorbate species with the cations in the glass (3), although there are numerous observations which indicate that the assumption of the ion-exchange as the only cause for adsorption is a gross oversimplification (*18*).

The adsorption of polyvalent metal ions on glass is a considerably more complex process. Such ions easily hydrolyze to give a variety of soluble complex species and, frequently, insoluble hydroxides at rather low pH values. Thus, an increase in pH not only affects the surface of the glass but also changes the entire composition of the adsorbate in solution. As a rule, the adsorption of polyvalent metal ions increases dramatically above a certain pH. In some cases, the adsorbed amount rises continuously with an increase in pH (*32*). In other cases, pronounced adsorption maxima were observed at some intermediate pH values. The position of the maximum varied with the element (*7, 8, 24, 26, 30, 31, 35, 36*). While there was evidence that this maximum was independent of the adsorbent (*4*), a big difference was found when glass and quartz were compared using the same counterion (*7, 8*). It is usually assumed that at very low pH, where hydrolysis is negligible, the polyvalent metal ions to be exchanged for cations in the glass are in competition with the high excess of hydrogen ions. The increase in adsorption with pH has been interpreted in various ways. In general, the hydrolysis of the adsorbing metal ions has been assumed responsible for the enhanced adsorption, although the use and the meaning of the term "hydrolysis" is not always consistent. In some instances this refers to the formation of soluble, while in other cases to insoluble products. The greater adsorptivity of soluble complex species was proposed by several investigators (*7, 8, 13, 14, 24, 30, 34, 36*). However, no one ever attempted to make a quantitative correlation between the composition of the adsorbate solution and the adsorbed quantities. In general it is believed that the adsorption depends on the charge of the ionic complexes and that lower charged or neutral sspecies adsorb less strongly (*34, 36*).

More frequently the dependence of adsorption upon pH was related to the formation of colloidal metal hydroxides. Again, some authors expressed the opinion that the formation of colloids promoted the adsorption (*31, 32*) while some others claimed the opposite (*7, 8, 24, 36*). However, it was generally agreed that when an adsorption maximum was observed as a function of the pH, the reduced adsorption at higher pH values was explained by the electrostatic repulsion between the colloid particles and glass surface bearing the same charge.

The entire picture is still more confusing because of the fact that several different types of colloids are distinguished—*i.e.*, "radiocolloids," "pseudo-colloids" (*7, 8, 28, 33*), and "true colloids." Radio-colloids refer to systems of radiotracers which appear to be in colloidal form although they are in concentrations well below their ionic solubility (*25, 26*). The term pseudo-colloid is used to describe the formation of a colloid system by the adsorption of a radioelement on solid impurities contained in the solution (7, 8, 29). While the nature of "radiocolloids" is still controversial (25, 27) the meaning of "pseudo-colloids" is completely obscure. One gets the feeling that the latter concept was introduced for lack of understanding of the complex process of adsorption of hydrolyzed species from aqueous solutions.

From the preceding survey it is apparent that further studies are needed to elucidate the mechanism of adsorption of hydrolyzable ions on various adsorbents, particularly on glass. Of special interest is the question whether an enhanced adsorption at higher pH is caused by soluble hydrolyzed species or by the formation of colloidal hydroxides. If the soluble complexes are responsible for the greater adsorptivity the relationship of the latter to the charge, size, shape, configuration, and ligand composition of the adsorbate species becomes pertinent. The answer to these problems is essential for the understanding of various surface phenomena in the presence of hydrolyzable ions such as sol stability, flotation, coprecipitation, adhesion, paper sizing, etc.

Several attempts have been made to correlate the adsorptivity of hydrolyzable cations to the composition of the species in aqueous solution (1, 2, 20). In particular, the adsorption of thorium on silver halides indicated a very close relationship between the change in the amount of thorium adsorbed and the concentration of the hydrolyzed soluble species in solution (19). The major difficulty in this type of work is the lack of quantitative data on the hydrolysis of various metal ions. The other uncertainty is with regard to the knowledge of the true surface area of the adsorbent in aqueous solution. This latter information is needed if surface coverages are to be evaluated.

At least some of these difficulties have been overcome in the work to be reported in this study, which deals with the adsorption of hafnium hydrolyzed species on powdered glass as a function of the acidity of the medium. The adsorption of hafnium species from aqueous solution has apparently never been investigated, yet this ion lends itself conveniently to studies of the problems discussed above. The chemistry of the hafnium ion in water is fairly well understood (23) and a suitable isotope, [181]Hf, is available for adsorption studies. What makes hafnium a particularly interesting system is the fact that it forms the entire series of hydrolyzed species: $Hf(OH)_n{}^{(4-n)+}$ where $n \leqslant 4$. At intermediate acidities (pH $>$ 4) the solutions of low concentrations contain only the neutral, soluble species $Hf(OH)_4$. It should be emphasized that there is a pH and a concentration range over which this species is present without simultaneous formation of hafnium hydroxide. Thus, it is possible to elucidate the effect of the ionic charge upon the adsorption of hydrolyzed species in systems void of colloidal hydroxides. The glass powder was used in order to have a sufficient surface area of adsorbent which can be determined with reasonable accuracy and which would not change appreciably upon dispersion in water. The preliminary work of the adsorption of hafnium on silver halide sols (21) could not be fully analyzed because the surface area of the adsorbent was not known.

Another advantage of hafnium is that, if the adsorption of the neutral species takes place, a close packed monolayer should eventually result owing to the absence of electrostatic repulsion. Knowing the surface area of the adsorbent this would enable one to evaluate the cross-sectional area of the hydrolyzed complex ion. This information has not been available.

Experimental

Materials. ^{181}Hf was employed in tracer concentrations and was obtained from Oak Ridge National Laboratories in the form of the chloride in approximately $1N$ HCl. The analysis of the gamma ray spectrum revealed that purification of the isotope was unnecessary. The solutions were prepared in the following manner. One portion of the acidified isotope solution was diluted to a desired volume with nitric acid giving a final concentration of $0.4N$ HNO_3. A second isotope solution was prepared in exactly the same manner except that the dilution was made with doubly distilled water, resulting in a final solution of which the pH was 3.5. A known quantity of these radioactive hafnium solutions was then added to a solution of stable hafnium tetrachloride to obtain a reasonable count rate over the concentration range studied. The pH of the final stock solution was either 2.0 or 3.1 depending on whether it was prepared with acid or water diluted tracer. A period of three days was allowed for equilibration between the solution and the container walls. All subsequent dilutions were prepared from these stock solutions. A need for the proper procedure of preparation of the labelled solutions of hydrolyzable metal ions was emphasized by several investigators (25, 26). Freshly prepared stock solutions, obtained as described above appeared to be homogeneous and free of colloidal precipitates. However, after prolonged storage traces of hafnium hydroxide were found. Such solutions were not used in experiments. Instead fresh stock solutions were prepared every few weeks.

Hafnium tetrachloride, nitric acid, and potassium hydroxide solutions were prepared using doubly-distilled water from an all borosilicate glass still. The chemicals were of the highest purity grade commercially available and were used without further purification.

The glass powder, which served as the adsorbent, was obtained from the Arthur S. Lapine Company in the form of spherical beads approximately 40 μ in diameter. The beads were washed with a large quantity of distilled water and dried in an oven at 120°C. before use in the adsorption experiments.

A sample of the glass powder was leached by refluxing with 15% HCl at 80°C. for approximately four days. After this the beads were washed thoroughly with distilled water and then heated in an oven at 380°C. for several hours.

The surface area of the adsorbent, cleaned by the first procedure, was measured by the B.E.T. method using water vapor as the adsorbate. Assuming the cross-sectional area of water to be 12.5 A.^{2} (5), three determinations resulted in a value of 0.80 ± 0.05 meter^2/gram. For comparison reasons, the geometric surface area was also determined from a histogram obtained by counting several hundred glass particles on microphotographs. This surface area was only 0.04 meter^2/gram. The difference between the two procedures employed indicates that the glass beads used exhibited significant surface roughness.

Method. All adsorption samples were prepared by weighing the glass powder on an analytical balance and by adding the appropriate amounts of radioactive hafnium, nitric acid or potassium hydroxide, and doubly-distilled water to give a constant final volume of 10 ml. The solutions were agitated by means of a magnetic stirrer for the desired periods of time whereupon the systems were centrifuged at 3,500 r.p.m., corresponding to 2.5 × 10^3 g, for 15 minutes in an IEC International Centrifuge. An aliquot of the supernatant solution was then removed for radioactive analysis. The remaining solution was used for pH measurements employing calibrated glass electrodes in a Beckman Model G pH meter. The glass beads from the same sample were washed twice with an inactive solution of hafnium of the same concentration as the mother liquor. The washings were removed and discarded. To determine the amount of hafnium adsorbed on the glass, 10 ml. of a 0.5N HNO$_3$ solution were added to the samples and stirred overnight after completion of the washing. This was necessary in order to maintain constant geometry conditions. Radioactive countings were performed with a Tracerlab Gamma Guard Fully Automatic Well Scintillation Console System.

Centrifugation was used to determine the precipitation limit of the hafnium solutions as a function of the pH. A series of solutions containing a constant amount of hafnium, to which sodium hydroxide was added in increasing amounts to vary the pH in small increments from system to system, was allowed to equilibrate for a certain period of time. The desired pH was adjusted automatically using a Radiometer Model PHM-28 pH meter, with glass electrodes. The solutions were then centrifuged at 3,500 r.p.m. for 1/2 hour. Five milliliter aliquots were drawn from the upper part of each of the solutions and the activity of this portion was compared with the activity of the remaining 5 ml. sample. From this the fraction of the centrifuged hafnium was determined. In certain cases, which will be discussed later on, it was necessary to ultracentrifuge the systems. The samples were prepared in an analogous manner and centrifuged in a Beckman Preparative Ultracentrifuge Model L-2 at 25,000 r.p.m., corresponding to 5.5 × 10^4 g, for one hour.

In all calculations, corrections were made for the adsorption on the test tubes. As a rule these corrections were small due to the small surface area of the test tubes in comparison to that of the glass powder used. All ^{181}Hf determinations were made on constant final volumes of 10 ml. to insure reproducible counting conditions.

Standard samples were prepared from the working solution of radioactive hafnium tetrachloride taking known volumes of this solution and diluting to a volume of 10 ml. These standards were always counted immediately following the adsorption samples to eliminate correction for decay losses. A linear relationship between the activity and the amount of hafnium dissolved was found over a concentration range of two orders of magnitude.

Experimental results showed that the combined activity obtained for the adsorbed hafnium and hafnium remaining in solution was within ±3% of the introduced activity. Table I gives some typical results indicating this good agreement.

Table I.

$HfCl_4$	Count Rate (c.p.m.) of the Adsorbed Amount	Count Rate (c.p.m.) of the Supernatant Solution	Total Count Rate (c.p.m.)	Introduced Count Rate (c.p.m.)	% Error
$1 \times 10^{-4}M$	442,137	723,982	1,166,119	1,136,364	+2.6
$1 \times 10^{-4}M$	495,500	636,286	1,131,786	1,136,364	−0.5
$5 \times 10^{-5}M$	332,049	248,720	580,769	568,729	+2.1
$5 \times 10^{-5}M$	281,611	284,244	565,855	568,729	−0.5
$2.5 \times 10^{-5}M$	213,658	3,204	216,860	210,713	+2.9
$2.5 \times 10^{-5}M$	169,867	39,528	209,395	210,713	−0.6

Results

Precipitation of Hafnium Hydroxide. In order to interpret the adsorption data it was necessary to determine the conditions which lead to the precipitation of hafnium hydroxide. It is not usually advisable to depend on the solubility product because the information on this quantity is often unreliable for hydroxides of polyvalent metal ions. In addition, "radiocolloids" may apparently form much below saturation conditions in radioactive isotope solutions. In the specific case of hafnium hydroxide only two measurements of the solubility seem to have been reported. According to Larson and Gammill (16) $K_s = [Hf(OH)_2^{2+}][OH^-]^2 = 4 \times 10^{-26}$ assuming the existence of only one hydrolyzed species $Hf(OH)_2^{2+}$. The second reported value is $K_{so} = [Hf^{4+}][OH^-]^4 = 3.7 \times 10^{-55}$ (15). If one uses the solubility data by Larson and Gammill (Ref. 16, Tables I and III) and takes into consideration all monomeric hafnium species (23) a K_{so} value of 4×10^{-58} is calculated.

Because of the inconsistency of these results, experiments were carried out to establish the precipitation boundaries, as described earlier. Figure 1 gives as an example four curves in which the fraction of hafnium removed by precipitation as hydroxide is plotted against the pH for four different concentrations of $HfCl_4$. Open and blackened symbols are for experiments in which systems were equilibrated before centrifugation 1 hour and 70 hours, respectively. In all cases insoluble precipitates are

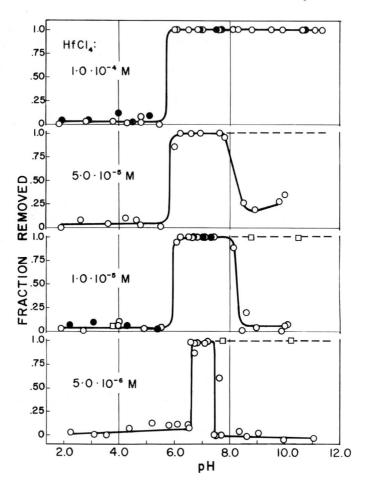

Figure 1. The fraction of hafnium tetrachloride removed from solution by centrifugation at 3,500 r.p.m. for 1/2 hour as a function of the pH. Open and blackened symbols represent centrifugation 1 hour and 70 hours after mixing the precipitating components, respectively. Squares represent the fraction of hafnium removed by ultracentrifugation at 25,000 r.p.m. for 1 hour and the corresponding dashed line represents the curves which would result from these studies

formed above a certain pH and this limit increases with a decrease in the salt concentration. In these examples the pH range for the onset of precipitation varies between 5.7 and 6.5. Above this pH the removal of hafnium by precipitation from the solution of hafnium chloride is nearly complete. However, except for the highest concentration of $HfCl_4$ the precipitation region is followed at still higher pH values by a region over which precipitation and settling of hafnium hydroxide does not take place under the experimental conditions employed. This could either be caused by the formation of soluble anionic complexes of hafnium or by extremely finely dispersed hydroxide. In order to distinguish between the two possibilities several samples in the second region of low fractions removed were ultracentrifuged as described. Results at pH values of 10.3 and 10.6 for $1 \times 10^{-5} M$ and $5 \times 10^{-6} M$ $HfCl_4$, respectively, showed that the hafnium is completely removed from solution indicating the presence of finely dispersed hydroxide. These points are indicated in the diagram as squares. If all the systems over the high pH range were ultracentrifuged the curves would look as indicated by the dashed lines. The original measurements at higher pH values obtained using the lower speed centrifuge are reported to show that erroneous conclusions may easily result owing to the stability of the extremely finely dispersed hafnium hydroxide. No attempt was made at this point to characterize this hydroxide sol.

Similar ultracentrifugation experiments were carried out with systems at pH values below the precipitation region. However, the results were identical to those obtained using the lower speed centrifuge as shown in Figure 1. It is therefore concluded that over the low pH range and the concentrations used the solutions of hafnium tetrachloride are void of colloidal particles.

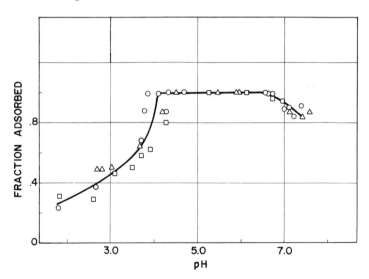

Figure 2. Fraction of hafnium adsorbed after 70 hours on 0.10 grams of glass beads as a function of the pH for a $5 \times 10^{-5} M$ hafnium tetrachloride solution

○ = *use of unacidified stock solution of $HfCl_4$*
△ = *use of acidified stock solution of $HfCl_4$*
□ = *use of acid treated glass beads and an acidified $HfCl_4$ solution*

Although data as reported in Figure 1 should lend themselves to the calculation of the solubility of hafnium hydroxide, neither the accuracy nor the amount of data is sufficient to make quantitative calculations.

Adsorption of Hydrolyzed Hafnium Species. Figure 2 shows the typical adsorption behavior of hafnium as a function of the pH. Indistinguishable adsorption curves are obtained for systems prepared from acidified (triangles) and unacidified (circles) stock solutions of ^{181}Hf. This indicates the absence of radiocolloids and implies homogeneous mixing of the isotope with the non-radioactive hafnium salt solution. It was also established that hafnium chloride solutions aged at room temperature for several weeks produced the same adsorption curve. Squares represent results obtained with the acid leached glass using the acidified isotope stock solution. The described treatment of the glass obviously did not change its adsorptive characteristics for the hafnium species.

Figure 3 gives the fraction of adsorbed hafnium as a function of the total original hafnium concentration for various equilibration times. The pH values were kept within a narrow range as indicated in the legend. The scatter in data at higher hafnium concentrations may be caused by small fluctuations in the pH. As can be seen in Figure 2, the adsorption is rather sensitive to the change of acidity in the vicinity of pH 3. Several

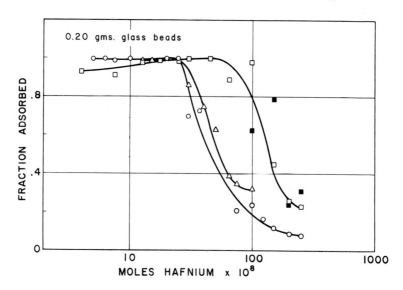

Figure 3. The fraction of hafnium tetrachloride adsorbed on 0.20 grams of glass beads as a function of the total hafnium concentration

○ = after 10 hours; pH 3.1-3.4
△ = after 20 hours; pH 3.3-3.5
□ = after 71 hours; pH 3.2-3.7
■ = after 119 hours; pH 2.9-3.0

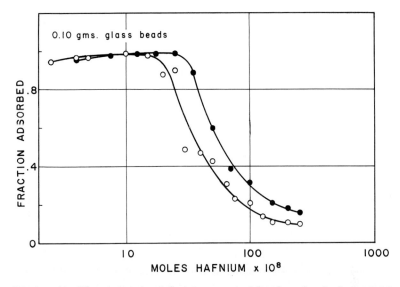

Figure 4. *The fraction of hafnium tetrachloride adsorbed on 0.10 grams of glass beads as a function of the total hafnium concentration*

○ = *after 20 hours of equilibration; pH 2.9-3.1*
● = *after 70 hours of equilibration; pH 3.0-3.2*

observations are noteworthy. Firstly, the adsorption seems to be a slow process: several days are needed until no further change in the adsorbed amount is detected. The results in Figure 3 indicate that approximately 70 hours are required to reach equilibrium saturation. A longer period of time (119 hours) had no further effect upon the amount of hafnium species adsorbed. Secondly, below a certain total concentration of $HfCl_4$, all of the hafnium is adsorbed. The concentration of hafnium above which only a fraction is adsorbed also shifts with time showing that the saturation of the surface is a time dependent process. These results enable one to calculate the amount of hafnium needed to saturate the surface of the glass.

Figure 4 is the same plot as in Figure 3 but for a smaller amount of glass. The difference in adsorbed quantities at 20 and 70 hours equilibration times is smaller than with the higher weight of the adsorbent.

The results in Figure 3 can be presented as the adsorbed amount of hafnium *vs.* the concentration of hafnium remaining in the solution. Both quantities were determined experimentally and neither was obtained by difference. This explains the apparently large scatter of results at higher hafnium concentrations. Data in Figure 5 show that above a rather low concentration of hafnium the adsorbed amount remains constant. However, this constant concentration increases with time until saturation is reached.

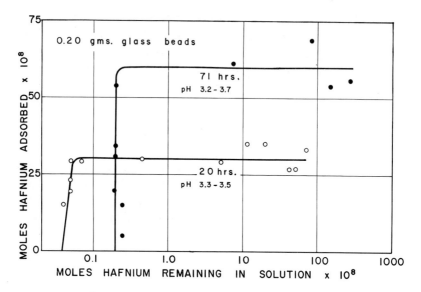

Figure 5. Adsorption of hafnium on 0.20 grams of glass beads after 20 hours (○) and 71 hours (●) as a function of the equilibrium solution concentration. The corresponding pH values are given in the diagram

The saturation values at equilibrium are given in Figure 6 for three different weights of glass beads. The quantity adsorbed at equilibrium

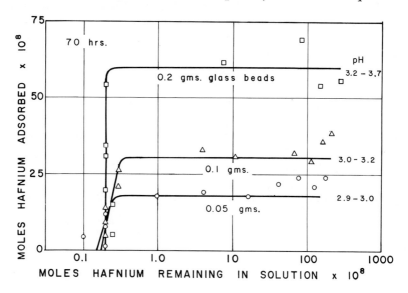

Figure 6. Adsorption of hafnium tetrachloride after 70 hours for 0.05 (○), 0.10 (△), and 0.20 (□) grams of glass beads as a function of the equilibrium concentration. The corresponding pH values are given in the diagram

would be expected to vary directly with the surface area, or in this case, the sample weight. The values obtained from the saturation levels were 18×10^{-8}, 31×10^{-8}, and 60×10^{-8} moles of hafnium adsorbed for 0.05, 0.10, and 0.20 grams of glass beads, respectively. Calculation of the adsorption ratios gave values of 1.2:2.1:4.0, in reasonably good agreement with the weight ratios of 1:2:4.

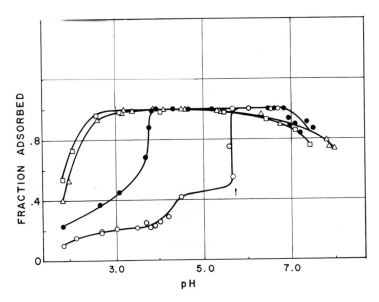

Figure 7. *Adsorption of hafnium on 0.10 grams of glass beads after 70 hours at four different $HfCl_4$ concentrations as a function of the pH*

$\square = HfCl_4$: $1.25 \times 10^{-5}M$
$\triangle = HfCl_4$: $2.5 \times 10^{-5}M$
$\bullet = HfCl_4$: $5.0 \times 10^{-5}M$
$\bigcirc = HfCl_4$: $1.5 \times 10^{-4}M$

Figure 7 shows the effect of pH upon the adsorption of hafnium species for various concentrations of the total hafnium chloride. It can be seen that except for the highest hafnium concentration and the lowest pH values all of the hafnium is adsorbed on the glass surface. These results would then indicate that adsorption of hafnium occurs even when the species carry no charge. It is interesting to note that the adsorbed amount begins to decrease somewhat at higher pH values. The sudden increase in the adsorption at a pH of 5.5 for the hafnium tetrachloride concentration of $1.5 \times 10^{-5}M$ is out of line with the rest of the curves. This is due to the formation of hafnium hydroxide. The beginning of precipitation as determined in Figure 1 is indicated by the arrow.

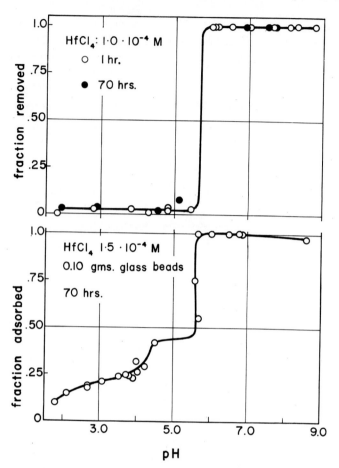

Figure 8. Centrifugation and adsorption of $HfCl_4$ as a function of the pH. Centrifugation (top curve) at 3,500 r.p.m. for 1/2 hour of a 1×10^{-4}M $HfCl_4$ solution 1 hour (○) and 70 hours (●) after mixing the reacting components. Adsorption (bottom curve) on 0.10 grams of glass beads at 70 hours for a 1.5×10^{-4}M $HfCl_4$ solution

Figure 8 illustrates the correlation between the apparent adsorption increase and the onset of precipitation of hafnium hydroxide.

Discussion

The results presented in Figures 7 and 8 and the centrifugation experiments in Figure 1 show that saturation adsorption of hafnium on glass is reached before the onset of hydroxide precipitation. Therefore, the adsorption results must be explained in terms of soluble hafnium species. Peshkova and Ang (23) have established that the hydrolysis of the hafnium(IV) ion leads to the formation of the following species with their corresponding hydrolysis constants given in parentheses:

$$Hf(OH)^{3+}(K_{1,1}=1.33), Hf(OH)_2^{2+}(K_{1,2}=0.59),$$
$$Hf(OH)_3^{+}(K_{1,3}=0.38), Hf(OH)_4(K_{1,4}=0.30),$$
$$Hf_3(OH)_4^{8+}(K_{3,4}=2.34 \times 10^4), \text{ and } Hf_4(OH)_8^{8+}(K_{4,8}=1.01 \times 10^8).$$

Using these constants the composition of the hafnium solutions was calculated as a function of the pH according to the following equation and the results are given in Table II.

$$[Hf]_{tot} = [Hf^{4+}] + K_{1,1}\frac{[Hf^{4+}]}{[H^+]} + K_{1,2}\frac{[Hf^{4+}]}{[H^+]^2} + K_{1,3}\frac{[Hf^{4+}]}{[H^+]^3} + K_{1,4}\frac{[Hf^{4+}]}{[H^+]^4} + 3K_{3,4}\frac{[Hf^{4+}]^3}{[H^+]^4} + 4K_{4,8}\frac{[Hf^{4+}]^4}{[H^+]^8}$$

Table II.

pH	% Hf^{4+}	% $Hf(OH)^{3+}$	% $Hf(OH)_2^{2+}$	% $Hf(OH)_3^{+}$	% $Hf(OH)_4$
2.0	0	0	0.02	1.25	98.73
3.0	0	0	0	0.13	99.87
4.0	0	0	0	0.01	99.99
5.0	0	0	0	0	100
6.0	0	0	0	0	100
7.0	0	0	0	0	100

In the investigated pH and concentration domain studied it was found that the amounts of the polymeric complex species were negligible and therefore only the mononuclear species are tabulated. This is also the reason why the percent composition as given in Table II is independent of the total concentration of the hafnium salt. It is obvious that at higher pH values the metal hydroxide will form if the initial concentration of the hafnium ion is sufficiently high.

Tyree (37) found evidence of highly polymerized hafnium species, but he studied solutions of high salt concentrations at elevated temperatures.

In view of the fact that no hafnium is present in the unhydrolyzed form under any of the conditions given in Table II, the ion exchange mechanism of adsorption by glass may be disregarded. This is substantiated by the fact that a different treatment of the glass had no effect upon its adsorption capacity for hafnium. This is similar to the results of Starik and Rozovskaya who found small effects upon the adsorptivity of hydrolyzed ions caused by glass modification even when drastic treatment such as fluoridization was employed. However, the adsorption of simple ions—e.g., Ag^+, decreased markedly on glass treated by fluoride (34).

It follows from Table II that hafnium adsorbs in the form of its hydrolyzed species. More importantly, the neutral $Hf(OH)_4$ species are strongly adsorbed. This contradicts directly statements made by several investigators (7, 8, 24, 36) that neutral species do not adsorb or adsorb less strongly than charged ions. It should be mentioned that their statements were based upon speculation rather than a quantitative analysis of the solution composition. The data presented here prove rather convincingly that the adsorptivity does not depend on the actual charge of the adsorbate species as has been assumed and therefore it cannot be considered strictly as an electrostatic process. It is the presence of the hydroxyl group in the hydrolyzed species that is responsible for its enhanced adsorptivity. The adsorption behavior of hafnium on glass is consistent with our earlier observations of the adsorption of this and

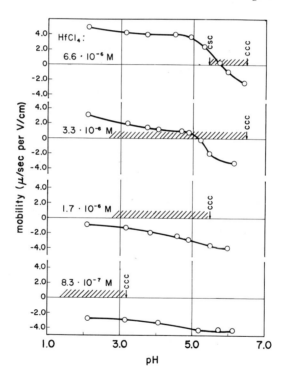

Figure 9. Mobilities of an aged silver iodide sol (AgI 1.0×10^{-4}M; $pI^- = 2.9$) in the presence of four different concentrations of $HfCl_4$ as a function of pH. Corresponding values for the critical coagulation and stabilization concentrations (c.c.c. and c.s.c. respectively) are indicated by arrows. Hatching represents the coagulation range

other ions, such as thorium, on silver halides (*19, 20, 21*). For example, at pH < 4.0 where thorium is essentially nonhydrolyzed the adsorption on a silver halide sol is negligible, but increases dramatically as soon as hydrolysis becomes significant (*see* Figure 9 and Reference *19*). This would then eliminate a possible alternate mechanism which would involve first the adsorption of the nonhydrolyzed metal ions with subsequent hydroxylation at the interface.

Below a certain pH the fraction of hafnium adsorbed decreases (Figures 2 and 7). This is most likely because of the competition of hydrogen ions for the adsorption sites.

It is also easily understood why in the case of hafnium the surface coverage increases as the pH becomes higher. Under these conditions the adsorbate consists only of neutral hafnium species. Since there is no lateral repulsion between them, the uncharged $Hf(OH)_4$ can adsorb until a close packed monolayer is formed. The experiments indicate no evidence of multilayer adsorption. Once saturation is reached the adsorbed amount remains constant regardless of the equilibration time or the concentration of the hafnium salt in solution.

It is of interest to calculate the area per molecule of the adsorbed species assuming that a monolayer is formed. In order to do so one needs exact information on the surface area of the adsorbent. Several attempts to calculate surface coverages in the adsorption of hydrolyzed metal ions have been reported (*4, 10, 11, 24, 27*). None of these results can be accepted as meaningful because the geometric surface area was always used in the calculations which is certainly smaller than the true surface area of the adsorbent. In addition, the adsorption was almost invariably measured on glass plates of several square centimeters in surface or on a few glass beads. Only in a few cases were glass powder or glass wool employed (*6, 7, 8, 26*).

The method of determining the "true" surface area is especially important in the case of glass as the adsorbent. Using five different samples of glass Jura (*12*) obtained the same surface area using nitrogen as the adsorbate but quite different results for each sample when water vapor was employed. Since in this work the adsorption of a solute from an aqueous solution was measured, water vapor was considered the most appropriate adsorbate to determine the surface available for adsorption. Calculation of the surface coverage by $Hf(OH)_4$ at pH 5 at equilibrium gives a value of 24 A.^{2} per hafnium species. It is interesting that this value is remarkably close to the area of 25 A.^{2} assigned to a water molecule chemisorbed on a silica surface (*5*). In fact the obtained surface area per molecule agrees quite well with the cross-sectional area of a tetrahedrally coordinated $Hf(OH)_4$ complex computed from the known bond lengths.

At lower pH values the areas per molecule at maximum coverage are larger. For example, at pH ~3 the calculated area per hafnium species is 42 A.^{2}. This is to be expected if one considers the hydrogen ion competition in the adsorption process.

These results are qualitatively supported by observations of adsorption of hafnium species on negative silver iodide sols (21). In the latter case it was also shown that at lower pH values charge reversal of the silver iodide takes place while at higher pH the particles remain negative. Figure 9 gives as an example four mobility curves as a function of pH for an aged silver iodide sol. This sol was prepared in an excess of potassium iodide and therefore in the absence of added hafnium salt the sol is negatively charged. According to Table II there should be no charged species present in hafnium salt solutions at pH $\geqslant$ 5, yet charge reversal does take place. This is easily understood because it can be calculated that the concentrations of charged hydrolyzed species required to neutralize or to reverse the charge are much below analytical detection limits. The concentrations of hydrolyzed species required to reverse the charge would be particularly low if the existence of highly aggregated charged species, as suggested by Tyree (37) is considered. The fact that at still higher pH values in the upper two curves the sol remains negative and that in the lower two curves reversal of charge does not take place at all indicate that the concentrations of the charged hydrolyzed species are extremely low and that the solution contains for all practical purposes only the neutral soluble species. The necessary consequence is that these species being uncharged although strongly adsorbed cannot cause charge reversal.

The glass beads used in this work were too large for electrophoretic measurements and therefore it was not possible to carry out mobility experiments with them.

Acknowledgment

We are greatly indebted to R. Sh. Mikhail, who carried out the B.E.T. measurements. We also acknowledge useful discussions with Donald Rosenthal.

This work was part of a Ph.D. thesis of one of the authors (L. J. S.), which was supported by a NASA Traineeship.

Literature Cited

(1) Abramson, M. B., Jaycock, M. J., Ottewill, R. H., *J. Chem. Soc.* **1964**, 5034.
(2) *Ibid.*, **1964**, 5041.
(3) Altug, I., Hair, M. L., *J. Phys. Chem.* **71**, 4260 (1967).

(4) Belloni, J., Haissinsky, M., Salama, H. N., *J. Phys. Chem.* **63**, 881 (1959).
(5) Brunauer, S., Kantro, D. L., Weise, C. H., *Can. J. Chem.* **34**, 1483 (1956).
(6) Deveaux, H., Aubel, E., *Compt. Rend.* **184**, 601 (1927).
(7) Grebenshchikova, V. I., Davydov, Yu. P., *Radiokhimiya* **3**, 155 (1961); Engl. trans. p. 167.
(8) *Ibid.*, **3**, 165 (1961); Engl. trans. p. 177.
(9) Haissinsky, M., Laflamme, Y., *J. Chim. Phys.* **1958**, 510.
(10) Haissinsky, M., Paiss, Y., *J. Chim. Phys.* **1959**, 915.
(11) Hensley, J. W., Long, A. O., Willard, J. E., *Ind. Eng. Chem.* **41**, 1415 (1949).
(12) Jura, G., "The Physical Chemistry of Surface Fibers," W. D. Harkins, Rheinhold Publishing Company, New York, 1952.
(13) King, E. L., *Natl. Nuclear Energy Ser.* **IV-14B**, 434 (1949).
(14) *Ibid.*, **MDDC-393**, 1946.
(15) Kovalenko, P. N., Bagdasarov, K. N., *Zhur. Neorg. Khim.* **7**, 1765 (1962); Engl. trans. p. 913.
(16) Larsen, E. M., Gammill, A. M., *J. Am. Chem. Soc.* **72**, 3615 (1950).
(17) Leng, H., *Sitzber. Akad. Wiss. Wien, Abt. IIa* **136**, 19 (1927).
(18) Long, A. O., Willard, J. E., *Ind. Eng. Chem.* **44**, 916 (1952).
(19) Matijević, E., in "Principles and Applications of Water Chemistry," p. 328, Wiley, New York, 1967.
(20) Matijević, E., Abramson, M. B., Ottewill, R. H., Schulz, K. F., Kerker, M., *J. Phys. Chem.* **65**, 1724 (1961).
(21) Matijević, E., Kratohvil, S., Stryker, L. J., *Discussions Faraday Soc.* **42**, 187 (1966).
(22) Nikitin, B. A., Vdovenko, V. M., *Trav. inst. etat radium (U.S.S.R.)* **3**, 256 (1937).
(23) Peshkova, V. M., Ang, P'êng, *Zhur. Neorg. Khim.* **7**, 2110 (1962); Engl. trans. p. 1091.
(24) Rydberg, J., Rydberg, B., *Svensk Kem. Tidskr.* **64**, 200 (1952).
(25) Schubert, J., Conn, E. E., *Nucleonics* **4**, 2 (1949).
(26) Schweitzer, G. K., Bishop, W. N., *J. Am. Chem. Soc.* **75**, 6630 (1953).
(27) Schweitzer, G. K., Jackson, M., *J. Chem. Educ.* **29**, 513 (1952).
(28) Starik, I. E., Ampelogova, N. I., *Radiokhimiya* **1**, 425 (1959).
(29) Starik, I. E., Ginzburg, F. L., Raevskii, V. N., *Radiokhimiya* **6**, 474 (1964); Engl. trans. p. 455.
(30) Starik, I. E., Kolyadin, L. B., *Zhur. Neorg. Khim.* **2**, 1432 (1957); Engl. trans. p. 349.
(31) Starik, I. E., Kositsyn, A. V., *Zhur. Neorg. Khim.* **2**, 444 (1957); Engl. trans. p. 332.
(32) *Ibid.*, **2**, 1171 (1957); Engl. trans. p. 286.
trans. p. 286.
(33) Starik, I. E., Rozovskaya, N. G., *Zhur. Neorg. Khim.* **1**, 598 (1956); Engl. trans. p. 267.
(34) Starik, I. E., Rozovskaya, N. G., *Radiokhimiya* **3**, 144 (1961); Engl. trans. p. 156.
(35) Starik, I. E., Sheidina, A. D., Il'menkova, L. I., *Radiokhimiya* **1**, 168 (1959); Engl. trans. p. 83.
(36) Starik, I. E., Skul'skii, I. A., Yurtov, A. I., *Radiokhimiya* **1**, 66 (1959); Engl. trans. p. 48.
(37) Tyree, S. Y., Jr., *Advan. Chem. Ser.* **67**, 183 (1967).

RECEIVED February 8, 1968. This work was supported by the Federal Water Pollution Control Administration Grant WP-00815.

6

The Adsorption of Aqueous Co(II) at the Silica–Water Interface

T. W. HEALY, R. O. JAMES, and R. COOPER

Department of Physical Chemistry, University of Melbourne, Parkville, Victoria 3052, Australia

> *The adsorption of Co(II) at the silica-water interface has been studied as a function of pH, ionic strength, and total Co(II) concentration. The adsorption data, together with electrophoretic mobility and coagulation data suggest that the free aquo Co(II) ion is not specifically adsorbed without participation of surface hydroxyls. Evidence for polymeric $Co(OH)_2$ at the quartz surface is presented together with evidence of mutual coagulation of the quartz and precipitated cobalt hydroxide.*

The adsorption of metal ions from aqueous solutions is a phenomenon of immediate interest to workers in many diverse disciplines. The incorporation of metals into geological sediments, removal of metal ions from industrial and civic effluent, interference of trace metal ions in analytical and electroanalytical chemistry, ore flotation, metallurgical leaching processes, and the stability of ceramic slips are all processes which are controlled to a large extent by interaction of metal ions with solid-liquid interfaces.

Recent studies indicate that the adsorption of metal ions is controlled only in part by the concentration of the free (aquo) metal ion; of considerable importance is the ability of hydroxo and other complex ions and molecules to adsorb. There have been two apparently divergent approaches to describe the role played by hydroxo metal complexes in adsorption at solid-aqueous electrolyte interfaces. Matijevic *et al.* (9) have proposed that specific hydrolysis products—*e.g.*, $Al_8(OH)_{20}^{4+}$ in the Al(III)-H_2O system, are responsible for extensive coagulation and charge reversal of hydrophobic colloids. It has also been demonstrated by Matijevic that the free (*aquo*) species of transition and other metal ions is frequently unable to reverse the charge of a sol whereas the hydrolysis products, often of lower charge per ion, can reverse the electrophoretic mobility of AgI sols (*in statu nascendi*) and rubber latex sols (*10*).

Alternatively, several workers have shown that not only is the soluble, zero-charged hydrolysis product considerably more surface active than the free (aquo) ion but also a polymeric charged or uncharged hydrolysis product may be formed at the solid-liquid interface at conditions well below saturation or precipitation in solution. Hall (*5*) has considered the coagulation of kaolinite by aluminum (III) and concluded that surface precipitates related to hydrated aluminum hydroxide control the adsorption-coagulation behavior. Similarly Healy and Jellett (*6*) have postulated that the polymeric, soluble, uncharged $Zn(OH)_2$ polymer can be nucleated catalytically at $ZnO-H_2O$ interfaces and will flocculate the colloidal ZnO *via* a bridging mechanism.

These two mechanisms, the one emphasizing the adsorption of specific often polynuclear hydrolysis products, the other emphasizing the role of polymeric species, are clearly not mutually exclusive; an earlier study on thorium (IV) adsorption suggested a combined mechanism (*1*).

The present study is on a system $Co(II)-H_2O-SiO_2$ for which it was expected that there would be minimal adsorption of polynuclear species of the metal ion but the possibility of surface catalysis to yield surface polymers of the hydroxide.

The oxide, α-quartz, was selected as the substrate for the present and continuing studies of metal ion adsorption. It is of considerable importance in several practical situations—*e.g.*, water purification and ore flotation—and has the important property that it is negatively charged over a wide pH range since its zero-point-of-charge (z.p.c.) is *circa* pH 2.

Experimental

The quartz dispersion was prepared by milling pure, acid leached, milky α-quartz specimens from Wattle Gully, Victoria, Australia, in a synthetic porcelain mill. The ball mill product was further cleaned by repeated washing-centrifugation with conductivity water. E.S.R. spectra of the as dried oxide powder itself and the dried powder when γ-irradiated in a ^{60}Co source, showed there were no paramagnetic impurities. This technique of analysis in which paramagnetic centers are generated around any multi-electron contaminant atom will be reported in detail shortly; it has proved useful in the detection of transition metal ion contaminants at the sub p.p.m. level on oxide surfaces. The quartz powder was found to have a B.E.T. surface area, based on krypton adsorption, of 5.4 meter^2/gram.

The metal ions were generated from their perchlorate salts which were either prepared from A.R. starting materials or purchased as A.R. chemicals. Solutions were freshly prepared for each experiment. Acid-base pH adjustment was made with A.R. perchloric acid and A.R. potassium hydroxide, respectively. Conductivity water was prepared by double distillation and during the course of the investigation had a conductance of 0.9 ± 0.1 micromhos cm.$^{-2}$. Water outside this upper limit was rejected. Cobalt (II) hydroxide, precipitated from perchlorate solution with KOH, was washed repeatedly with conductivity water.

While metal ions may be said to adsorb strongly at the oxide-water interface, adsorption studies are hampered by the need to determine metal ions at the sub micromolar level. Again the similarity between silica-water and glass-water interfaces means that the metal ion species will adsorb on all glass surfaces. For determination of the adsorption isotherms of Co(II), radioactive tracer techniques were employed using the ^{60}Co isotope. For this γ-emitting isotope liquid Geiger-Muller tubes were found to be satisfactory. At all times "blank" runs were conducted to correct for adsorption on the glass vessels.

Equilibration of the silica-water suspension with the metal ion was conducted in a thermostatted vessel similar to that described previously (15). Holes in the loose fitting lid were provided for pH electrodes,

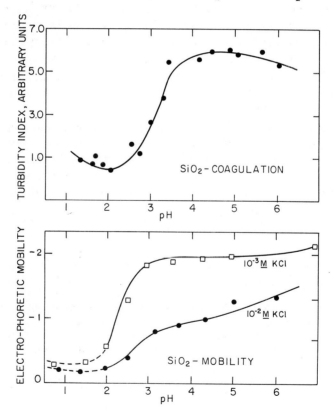

Figure 1. The coagulation and electrophoretic mobility (microns sec.$^{-1}$/volt cm.$^{-1}$) behavior of quartz in electrolyte solutions as a function of pH

nitrogen gas, and for capillary plastic tubing for sample removal and acid/base addition from an automatic pH-stat facility. It was found necessary to allow from 3 to 12 hours for equilibrium to be established. This kinetic effect is currently being investigated in more detail.

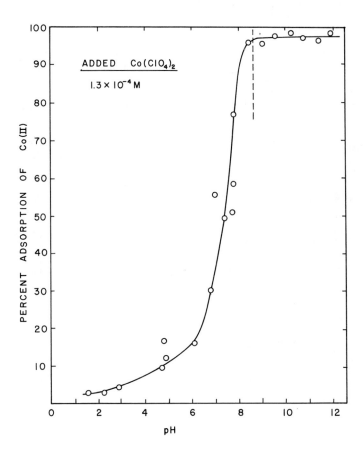

Figure 2. Percent adsorption on quartz of cobalt (II) as a function of pH for 1.3×10^{-4}M total $Co(ClO_4)_2$

Electrophoretic mobilities of the quartz particles in cobalt (II) perchlorate solutions were determined with a calibrated Zeta-Meter apparatus. Coagulation sedimentation behavior was followed using a stop-flow type apparatus. The dispersion is pumped in a closed loop from an equilibration vessel through an optical cell located in the sample compartment of a recording spectrophotometer. From the optical density-time curve obtained from the time the pump is switched off, the turbidity index (in arbitrary units) is obtained as the slope of the curve at zero time.

Results

The general surface properties of quartz are shown in Figure 1 in which the electrophoretic mobility and turbidity of quartz in simple electrolytes is plotted as a function of pH. The dotted lines shown in Figure 1 represent a region of ionic strength for which inaccuracies in mobility determination prevent the accumulation of meaningful data. Although the approach to the z.p.c. of quartz is obscured by this effect, the turbidity data confirm that the z.p.c. is at pH 1.8–2.0 in agreement with other workers (13).

The adsorption of cobalt (II) at $1.3 \times 10^{-4}M$ Co(ClO_4)$_2$ is shown in Figure 2 in the pH range from 1.7 to 12.0. This form of plot, percent adsorption vs. pH or concentration while useful for demonstrating the dramatic increase in adsorption over a narrow pH or concentration range, is however of limited theoretical value. In Figure 3 the cobalt (II) adsorption data are therefore redrawn as log (adsorption density) vs. pH. The vertical dashed lines in Figures 2 and 3 represent the minimum pH for precipitation of $1.3 \times 10^{-4}M$ Co(II) in the absence of adsorption. The plateau of Figure 3 therefore represents adsorbed and precipitated cobalt.

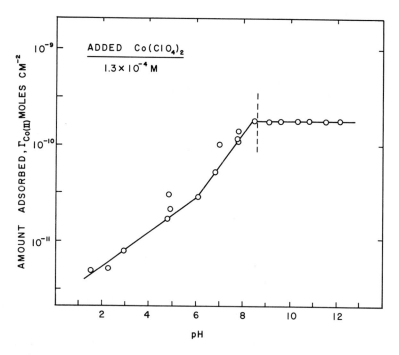

Figure 3. Adsorption isotherm of cobalt (II) on quartz as a function of pH and at $1.3 \times 10^{-4}M$ total Co(ClO_4)$_2$

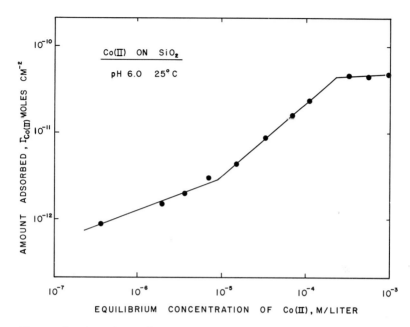

Figure 4. Complete adsorption isotherm of Co(II) on quartz at pH 6.0 and 25°C.

The adsorption of Co(II) at pH 6.0 is shown in Figure 4, again plotted on a log-log scale. Since cobalt (II) hydroxide does not precipitate at pH 6 at less than 1M total cobalt, the plateau attained represents saturation adsorption without interference by precipitation.

The variation with pH of the electrophoretic mobility of quartz in $10^{-4}M$ cobalt (II) perchlorate and a comparison of the mobility of quartz in $10^{-4}M$ KCl is shown in Figure 5. Included in Figure 5 is the variation with pH of electrophoretic mobility of precipitated cobalt (II) hydroxide. It can be seen that the silica surface with adsorbed Co(II) acts as cobalt (II) hydroxide for pH values above 8.0. The turbidity *vs.* pH behavior at $10^{-4}M$ Co(ClO$_4$)$_2$ is shown in Figure 6. The two curves represent the behavior for increasing and decreasing pH and within experimental error the curves superimpose.

Discussion

The most general feature of the adsorption behavior of metal ions at solid-aqueous solution interfaces is the abrupt rise in adsorption over a narrow pH range. This has been illustrated, for example, for manganese adsorption on glass (2), cobalt on hydrous ferric oxide (8), manganese on hydrous manganese oxide (12), protactinium on glass (14), and

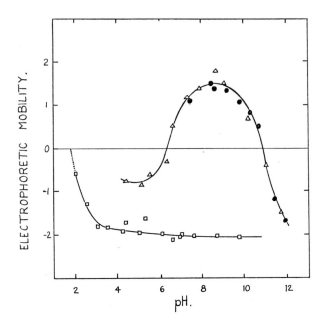

Figure 5. Electrophoretic mobility (microns sec.$^{-1}$/ volt cm.$^{-1}$) of SiO_2 in different electrolyte solutions together with the electrophoretic mobility of cobalt (II) hydroxide as a function of pH

—△— SiO_2, 10^{-4}M Co(II)
—●— $Co(OH)_2$, 10^{-4}M KCl

thorium on silver iodide (11). If we compare the pH range at which increased adsorption occurs with the properties of the solution in the same pH range there is often a striking parallel observed. For example, Matijevic et al. (11) have shown that the increase in adsorption of thorium occurs over the same pH range where the ratio $[Th(OH)^{3+}]/[Th(NO_3)_4]$ also increases abruptly.

We can generalize by noting that there will be increased adsorption at the p*K_1 of the aquo complex—i.e., *K_1 for

$$M(H_2O)_6^{n+} + H_2O \rightleftharpoons M(H_2O)_5(OH)^{(n-1)+} + H_3O^+ \tag{1}$$

Thus, for thorium (IV) the adsorption increase occurs at pH 3.82 (13). However, for Co(II) with a p*K_1 value of 9.8, the adsorption increase does not correspond to p*K_1 but is displaced to a lower p Hvalue.

Comparison of Figures 2 and 5 shows that at the pH range 6.5 to 7.5 where increased adsorption occurs, simultaneous reversal of charge and equivalent coagulation are both observed. This suggests that strong adsorption of a cationic cobalt (II) species is occurring. The principal cobalt (II) solution species reported in the literature (3) are represented in Figure 7 for the following self-consistent set of stability constants

$$Co(OH)_2 \rightleftharpoons Co^{2+} + 2OH^- \quad ; \log K_{s0} = -14.8 \quad (2)$$

$$Co^{2+} + OH^- \rightleftharpoons CoOH^+ \quad ; \log K_1 = 4.2 \quad (3)$$

$$Co^{2+} + H_2O \rightleftharpoons CoOH^+ + H^+ \quad ; \log {}^*K_1 = -9.8 \quad (4)$$

$$Co(OH)_2 \rightleftharpoons Co(OH)_2 aq. \quad ; \log K_{s2} = -6.40 \quad (5)$$

$$Co(OH)_2 + OH^- \rightleftharpoons Co(OH)_3^- ; \log K_{s3} = -5.10 \quad (6)$$

Here, $Co(OH)_2$ represents the solid hydroxide. The solution data show that at pH values of 7.5 and 6.5 the dominant cobalt (II) species is the free (aquo) ion by factors of 100 and 1000 respectively. It is therefore highly unlikely that the coagulation at pH 6.5–7.5 and $10^{-4}M$ Co(II) and the reversal of charge can be caused by the free $CoOH^+$ species. If it is caused by polynuclear charged species then the log–linear relationship (9) between the critical coagulation concentration and the valence of the coagulating ion would require a polynuclear species to have a charge of +5 or +6. Such a species has not been identified. (It is of interest to note that if this species did exist it would have to be a compact ion,

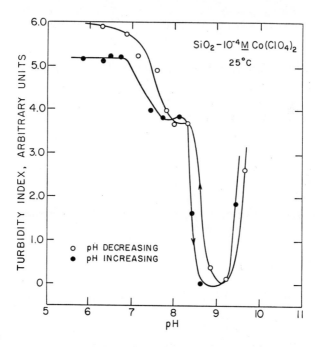

Figure 6. Coagulation of silica in aqueous $10^{-4}M$ Co(II) solution as a function of pH. Increase in the turbidity index indicates an increase in dispersion

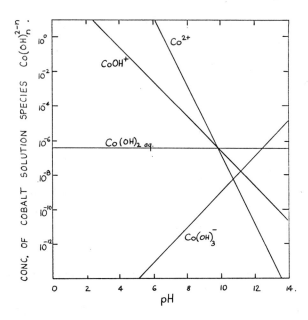

Figure 7. A summary of the variation with pH of the concentration of known solution species of cobalt (II) in aqueous solution in equilibrium with solid Co(II) hydroxide at 25°C.

since a linear hydroxyl-bridged octahedral Co(II) polymeric ion has a net charge of $+2$.) If we therefore reject polynuclear polyvalent Co(II) species, the pH 6.5–7.5 coagulation/charge reversal must be caused by the free (aquo) ion. However, since the silica surface charge is not increasing to any significant extent from pH 4.0 onwards, it is difficult to explain the drastic increase in adsorption shown in Figure 2 if the free Co^{2+} ion alone were the active species. However, if there is some cooperative adsorption of Co^{2+} with either bound or free hydroxide ions then an abrupt increase in adsorption is possible; remembering that the hydroxide ion concentration in the interface is greater than in solution, then specific Co^{2+} interaction with surface hydroxide ions occurs as observed at a lower pH than in bulk.

We can substantiate this model further by considering the adsorption isotherm shown in Figure 4 in terms of the Stern-Grahame treatment of the double layer (4). At concentrations below $10^{-5}M$ the slope of the log (adsorption)–log (equilibrium concn.) isotherm has a slope of approximately 0.5. This suggests that the Co^{2+} ion is acting as a counterion in the diffuse layer where its adsorption is described by a simple Gouy-Chapman expression. It is not specifically adsorbed and cannot reverse the sign of the surface charge. Above $10^{-5}M$ at pH 6.0 the slope is approximately 1.0 indicating that specific adsorption is occurring and that the adsorption is given by an equation of the form

$$\Gamma_i = k_i C_i \exp.(-W_\delta/kt) \qquad (7)$$

where Γ_i is the adsorption density in moles/cm.^{2}
 C_i is the equilibrium concentration in moles/cc.
 k_i is a constant, and,
 W_δ the free energy of adsorption of an ion i at a plane δ in the double layer is, for cation adsorption, given by,

$$W_\delta^+ = Z_+ e(\psi_\delta + \phi_+) \qquad (8)$$

where Z_+ is the valence of the ion
 e is the electronic charge
 ψ_δ is the potential at the plane of adsorption
and ϕ_+ is the specific adsorption potential.

In terms of Equations 7 and 8 the change in adsorption behavior at pH 6 and $10^{-5}M$ Co(II) shown in Figure 4 can be described as being caused by the operation of a specific adsorption potential (ϕ_+ cal./mole). Since the hydrolysis products are unlikely to contribute a sufficiently large potential to account for the increased adsorption, it must be concluded that above $10^{-5}M$ at pH 6, and above pH 6.5–7.5 at $10^{-4}M$ Co(II), the Co^{2+} ion is specifically adsorbed and located within the Stern plane. It is probable that these conditions correspond to the fact that an activity ratio of Co^{2+} and surface O$^-$ or OH$^-$ sites has been exceeded.

The plateau shown in Figure 3 corresponding to adsorbed and precipitated cobalt (II) is not amenable to analysis. However, the plateau of Figure 4 must correspond to saturation, or "monolayer" adsorption. Depending on whether the Co(II) ion is unhydrated, hydrated to the extent of one sheath of water or hydrated to the extent of two layers of water then the fractions of a close packed monolayer for these three cases are 0.01, 0.16, and 0.5 respectively. Clearly this low coverage means that there must be considerable repulsion between adsorbed Co(II) ions on the adsorbed layer leading to a sparse population of the surface.

The extensive coagulation at pH 8.5–9.0 for $10^{-4}M$ Co(II) occurs when the mobility of the SiO$_2$ in Co(II) perchlorate is large and positive. If we consider that the coagulation shown in Figure 6 at pH 6.5–7.5 is caused by Co^{2+} co-ordination with surface anion groups on a 1 to 1 basis then the coagulation at pH 8.5–9.0 would perhaps correspond to coordination on a 1 to 2 basis—*i.e.*, nucleation, probably in polymeric form of the cobalt hydroxide. This is further confirmed by the fact that the zero mobility at pH 10.6 shown in Figure 5 corresponds almost exactly to the z.p.c. of cobalt hydroxide precipitated in the absence of silica. The surface of silica with adsorbed cobalt at pH values above pH 8 is essentially a $Co(OH)_2$ surface. For the $10^{-3}M$ Co(II) case shown in Figure 6 there is also the possibility of coagulation—*i.e.*, mutual or hetero-coagulation—between precipitated cobalt hydroxide and the silica itself, since in this pH region the two solids are oppositely charged. As in the previous study (*6*), flocculation (*via* polymer bridges) of the silica by polymeric surface cobalt species may also be occurring. This effect is currently being examined in more detail by spectroscopic techniques.

Summary and Conclusions

The adsorption of Co(II) at the silica-water interface can be separated into three parts:

A. At low pH (<6.0) and low concentrations ($<10^{-5}M$) the active species is the Co^{2+} ion adsorbing as a nonspecifically adsorbing ion in the diffuse layer.

B. At pH-concentration conditions above 6.5 but well below precipitation, Co^{2+} is specifically adsorbed into the Stern layer, with coagulation and charge reversal accompanying this process. The surface interaction is probably of the form

$$-Si-O^{-}-\underset{\underset{H_2O}{\overset{H_2O}{|}}}{\overset{\overset{H_2O}{|}\overset{H_2O}{\diagup}}{Co}}{}^{2+}-H_2O \quad \text{or} \quad -Si-OH-\underset{\underset{H_2O}{\overset{H_2O}{|}}}{\overset{\overset{H_2O}{|}\overset{H_2O}{\diagup}}{Co}}{}^{2+}-H_2O$$

C. At pH/concentration conditions just below saturation the adsorbed species is probably a polymeric form of cobalt (II) hydroxide. At higher pH values the nucleation of cobalt (II) hydroxide is completed and the silica with adsorbed cobalt (II) behaves as cobalt (II) hydroxide. Some mutual coagulation between SiO_2 and precipitated $Co(OH)_2$ may occur for the higher Co(II) concentrations.

Acknowledgments

This work was supported by a research grant from the Australian Mineral Industries Research Association. This support is gratefully acknowledged. Financial support in the form of a University of Melbourne Research Grant to R. O. J. is also acknowledged. Assistance from Dianne John in the experimental program is gratefully acknowledged.

Literature Cited

(1) Abramson, M. B. *et al.*, *J. Chem. Soc.* **1964**, 5041.
(2) Benes, P., Garba, A., *Radiochim. Acta* **5**, 99 (1966).
(3) Bjerrum, J., Schwarzenback, G., Sillen, L. G., "Stability Constants," London Chem Soc., London, England, 1956.
(4) Grahame, D. C., *Chem. Rev.* **41**, 441 (1947).
(5) Hall, E. S., *Discussions Faraday Soc.* **42**, 197 (1966).
(6) Healy, T. W., Jellett, V. R., *J. Colloid Interface Sci.* **24**, 41 (1967).
(7) Hunt, J. P., "Metal Ions in Aqueous Solution," Benjamin, New York, 1965.
(8) Kurbatov, M. H., Wood, G. B., Kurbatov, J. D., *J. Phys. Chem.* **55**, 1170 (1951).
(9) Matijevic, E., Kratiovil, S., Stryker, L. J., *Discussions Faraday Soc.* **42**, 187 (1966).
(10) Matijevic, E., "Principles and Applications of Water Chemistry," p. 328, S. D. Faust, J. V. Hunter, eds., Wiley & Sons, New York, 1967.
(11) Matijevic, E. *et al.*, *J. Phys. Chem.* **65**, 1724 (1961).
(12) Morgan, J. J., Stumm, W., *J. Colloid Sci.* **19**, 347 (1964).
(13) Parks, G. A., *Chem. Rev.* **65**, 177 (1965).
(14) Sakanoue, M., Takagi, T., Maeda, M., *Radiochim. Acta* **5**, 79 (1966).
(15) Yopps, J. A., Fuerstenau, D. W., *J. Colloid Sci.* **19**, 61 (1964).

RECEIVED November 24, 1967.

7

The Adsorption of Aqueous Metal on Colloidal Hydrous Manganese Oxide

D. J. MURRAY[1]

Department of Mineral Technology, University of California, Berkeley, Calif.

T. W. HEALY

Department of Chemistry, University of Melbourne, Victoria, Australia

D. W. FUERSTENAU

Department of Mineral Technology, College of Engineering, University of California, Berkeley, Calif.

> *The adsorption of Groups I and II cations on manganese (II) manganite in aqueous suspension has been found to be strongly dependent on pH at low concentrations but is independent of pH at higher concentrations. In dilute solution, adsorption must then occur as counter ions in the diffuse double layer, whereas the extensive uptake at higher concentrations indicates that these ions are incorporated into the disordered layer of the manganite lattice. In addition, adsorption studies with the transition metal ions show that Ni^{2+}, Cu^{2+} and especially Co^{+2} exhibit marked specific adsorption as evidenced by the finite adsorption of these ions at the zero-point-of-charge.*

The ability of colloidal, hydrous manganese oxides to adsorb large quantities of aqueous metal ions has been a continuing subject of study since van Bemmelen's work of 1881 (5). While certain aspects of the subject have been well established—e.g., hydrogen ions are released (or hydroxide ions adsorbed) in proportion to the quantity of metal ion adsorbed (11)—there is still confusion as to the details of the mechanism of ion adsorption.

[1] Presently in the U. S. Army.

The recent comprehensive study of the general colloid-chemical properties of a hydrous manganese dioxide by Morgan and Stumm (10) did much to clarify many aspects of the adsorption process. A tentative assignment of the zero-point-of-charge (z.p.c.) of the essentially amorphous hydrous MnO_2 of less than pH 3 was made by these workers. Sorption of Mn^{2+} on MnO_2 about the z.p.c. was interpreted either as surface complex formation or as ion exchange. Use of the mass action analysis of Kurbatov (9), was shown by Morgan and Stumm (10) to be of limited quantitative significance for the extensive adsorption of cations on MnO_2.

Following Morgan and Stumm (10), the surface properties of five manganese oxides were examined by Healy, Herring, and Fuerstenau (7). Zero-points-of-charge of each oxide were determined by electrophoresis and coagulation techniques and found to range in a predictable manner from pH 1.5 for δ-MnO_2, pH 1.8 for Mn(II) manganite, 4.5 for α-MnO_2, pH 5.5 for γ-MnO_2, to pH 7.3 for β-MnO_2.

The present study was initiated in order to obtain quantitative data on the relative adsorption potentials of metal ions in the region of the z.p.c. of hydrous manganese oxide. This information is of considerable importance in a variety of practical phenomena ranging from the mechanism of trace metal inclusion in ocean-floor manganese nodules and pisolitic manganese ores to the sorption behavior of manganese precipitates in natural water and waste systems.

Experimental

A study of the zero-point-of-charge of various manganese dioxide preparations has been reported previously (7). The oxide selected for the present study was manganese (II) manganite or "10A. manganite" having a B.E.T. surface area of 70 meter^2/gram. The washed oxide was kept under twice distilled water at all times. Reagent grade nitrates of lithium, potassium, sodium, calcium, barium, copper (II), nickel (II), and cobalt (II) were used without further purification. The pH was adjusted with reagent grade nitric acid.

Conditioning of the manganese oxide suspension with each cation was conducted in a thermostatted cell (25° ± 0.05°C.) described previously (13). Analyses of residual lithium, potassium, sodium, calcium, and barium were obtained by standard flame photometry techniques on a Beckman DU-2 spectrophotometer with flame attachment. Analyses of copper, nickel, and cobalt were conducted on a Sargent Model XR recording polarograph. Samples for analysis were removed upon equilibration of the system, the solid centrifuged off and analytical concentrations determined from calibration curves. In contrast to Morgan and Stumm (10) who report fairly rapid equilibration, final attainment of equilibrium at constant pH, for example, upon addition of metal ions was often very slow, in some cases of the order of several hours.

Results and Discussion

The adsorption of metal ions in such systems can be treated in one of three ways, depending on whether the adsorbing ion is

(a) *Potential determining*—i.e., that conjugate pair of ions which define the total double layer potential, or

(b) *Surface inactive*—i.e., an ion which does not enter the Stern layer and which cannot reverse the potential of the outer Helmholtz plane of the electrical double layer, or

(c) *Surface active*, or specifically adsorbing—i.e., an ion which can enter the Stern layer and can reverse the sign of the potential of the outer Helmholtz plane.

Previous studies (7, 10) on manganese oxides have established that H^+ and OH^- are potential-determining for the MnO_2 series; for 10A. manganite, electrophoresis and coagulation measurements (Figure 2) both gave a value of pH 1.8 ± 0.5 for the z.p.c. (7).

The nature of adsorption of Group I ions on the oxide is illustrated in Figure 1 in which the adsorption isotherms for potassium ions on MnO_2 are given as a function of pH. Similar behavior was observed for the adsorption of sodium ions.

The pH dependence of adsorption at lower concentrations suggests that the adsorption occurs primarily as counterions in the diffuse part

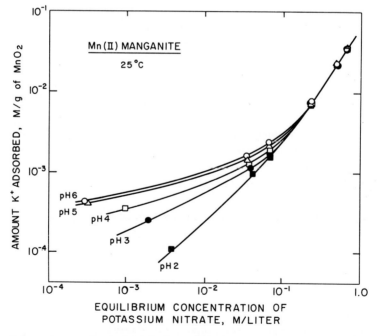

Figure 1. Isotherms for the adsorption of potassium ion on manganese (II) manganite at 25°C. at various pH values

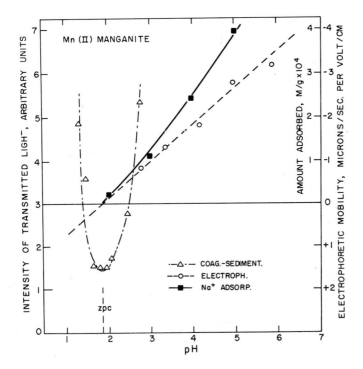

Figure 2. The zero-point-of-charge of manganese (II) manganite as determined by the adsorption of sodium ions, electrophoresis, and coagulation

of the double layer. As shown in Figure 2, the adsorption of Group I cations offers further evidence for the assignment of pH 1.8 as the z.p.c. of Mn(II) manganite. The reduction of the adsorption essentially to zero at the z.p.c. strongly supports the hypothesis that at lower concentrations these ions act as an indifferent, nonspecific electrolyte.

It is interesting to note that at higher concentrations the adsorption is essentially independent of pH. The reasons for this phenomenon are not clear but it seems likely that metal ions are being incorporated into the MnO_2 lattice. This oxide is one of the general class of "manganites" which consist of ordered layers of MnO_2 alternating with disordered layers of metal ions coordinated by H_2O, OH^- and other anions. In the 10A.-manganite the ordered layers, containing Mn^{4+} in six-fold coordination with O^{2-}, are separated by a 10A. disordered layer containing Mn^{2+} coordinated with O^{2-}, H_2O, and OH^- (2, 12).

That there may be metal ion uptake in the interlayer region is consistent with the above crystal structure; experimental confirmation of this process is provided by the work of Wadsley (12) and Buser et al. (2). However, these and other workers did not isolate the role of pH in such processes and the present study was therefore undertaken to clarify the role of pH in the adsorption-uptake process for this system.

A series of experiments was run at pH 4 to study the adsorption of Ca^{2+} and Ba^{2+}, and within the experimental error it was found that Ca^{2+} and Ba^{2+} adsorbed similarly at all concentrations. On the basis of ionic radii (Ca^{2+} 0.99A., Ba^{2+} 1,35A.) some difference would be expected. In Figure 3, the adsorption of Li^+, Na^+, and K^+ at pH 4 can be compared with the adsorption of Ba^{2+} and Ca^{2+} for the same metal ion concentration of 5×10^{-4} mole per liter. The placement of the Ca^{2+}, Ba^{2+} data relative to Na^+ and K^+ suggest that a small specific adsorption potential may be operative. For Ca^{2+} adsorbing on a carboxylate surface, Davies and Rideal (4) report specific adsorption potentials of the order of 4 kT compared with about 1.0 kT for Na^+ on the same substrate.

The adsorption isotherms for Ni^{2+}, Cu^{2+}, and Co^{2+} are illustrated in Figure 3. While the extent of Ni^{2+} adsorption at around pH 4 is similar to

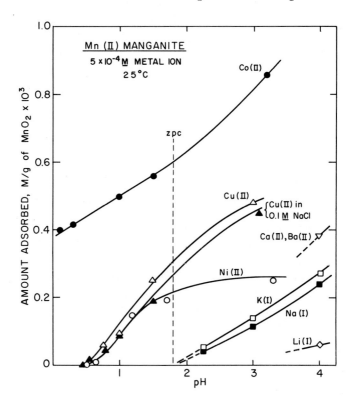

Figure 3. The variation of the adsorption of cobalt, copper, nickel, potassium, and sodium ions at 5×10^{-4}M concentration and 25°C. as a function of pH. Single data points for calcium, barium, and lithium ions are included

that of the Group II ions, Cu^{2+} and particularly Co^{2+} adsorb much more strongly. Furthermore, whereas Na^+ and K^+ can be desorbed by lowering the pH to that of the z.p.c., Ni^{2+}, Cu^{2+}, and especially Co^{2+} are not desorbed even from strongly positively charged surfaces. Such behavior can only be attributed to the existence of a relatively large specific adsorption potential.

Following Grahame (6) we can express the specific adsorption of cations as

$$\Gamma_+ = 2rC \exp \frac{-z_+ e(\psi_\delta + \phi_+)}{kT} \qquad (1)$$

where Γ_+ is the adsorption density in the Stern plane, r the radius of the adsorbed ion, C the equilibrium concentration, and ϕ_+ is the specific adsorption potential of the ion.

To evaluate ϕ_+ for each metal ion, values of ψ_δ are required at each concentration. While this can often be evaluated from electrophoretic mobility data, the high ionic strengths—i.e., pH < 2—preclude meaningful measurement of mobilities. However, it can be seen that when ψ_δ and ϕ_+ are equal and opposite then adsorption is reduced to zero. The adsorption of Na^+ is reduced to zero at the z.p.c. since, in this case, ϕ_+ is negligibly small. With Ni^{2+} and Cu^{2+} the pH must be reduced—i.e., ψ_δ made more positive—by 1.3 pH units to effect zero adsorption. Since near the z.p.c. ψ_δ and ψ_0, the total double layer potential, are approximately equal and given by the Nernst Equation, then

$$\psi_\delta \cong \psi_0 = \frac{RT}{F} \ln \frac{a_{H^+}}{a^\circ_{H^+}} \qquad (2)$$

where $a^\circ_{H^+}$ is the activity of H^+ ions at the z.p.c. and a_{H^+} is the activity at any given pH. Thus, for Ni^{2+} and Cu^{2+} the value of ϕ_+, the specific adsorption potential is approximately -80 mv. or -1.9 kcal./mole. It is interesting to note that addition of $10^{-1}M$ NaCl to the Cu^{2+} solution had little effect on the Cu^{2+} adsorption isotherm. This is in keeping with the almost zero value of the specific adsorption potential for Na^+.

Within the range of the present experiments the adsorption of Co^{2+} was never reduced to zero and so evaluation of the specific adsorption potential for Co^{2+} is not possible. It is significantly larger than that for Cu^{2+} and Ni^{2+} and is probably of the order of -5 kcal./mole.

It is surprising that the adsorption of Co^{2+} is anomalous when compared with other transition metal ions in that Co(II) and Ni(II) and to a lesser extent Cu(II) are very similar chemically. It is known (8) that certain metal ions can be catalytically oxidized on metal electrodes. It may therefore be possible for Co^{2+} in the present system to be oxidized to Co^{3+} in the region of the MnO_2-water interface. Although this suggestion has been made previously (1), it is still not possible to describe the increased adsorption caused by such oxidation quantitatively in terms of a separate specific adsorption potential, since $E°$ values for surface redox reactions are unknown. Similar oxidations for Cu^{2+} and Ni^{2+} are not expected since the higher valence states do not normally exist. In view of the similar chemistry of Co(II) and Ni(II) yet vastly different adsorption properties, the oxidation of Co^{2+} and Co^{3+} at the MnO_2–water interface appears to be the most likely mechanism to account for the adsorption.

Summary and Conclusions

The adsorption of Group I and II cations on manganese (II) manganite has been shown to take place in the diffuse part of the electrical double layer at low concentrations of the ions. At higher concentrations, the lack of pH dependence suggests that some other mechanism must be operating. It is considered that under these conditions, the ions are being incorporated into the disordered layer of the manganite lattice.

The transition metal ions Ni^{2+}, Cu^{2+}, and especially Co^{2+} exhibit marked specific adsorption as evidenced by the finite adsorption of these ions at the z.p.c. The apparently anomalous behavior of Co^{2+} which is adsorbed much more strongly than Cu^{2+} and Ni^{2+} might be attributed to surface oxidation of Co^{2+} to Co^{3+}.

Acknowledgments

The authors wish to thank the National Science Foundation for support of this research in the Department of Mineral Technology, University of California, Berkeley. T. W. Healy acknowledges support by a Queen Elizabeth Fellowship at the University of Melbourne for part of this work.

Literature Cited

(1) Burns, R. G., *Nature* **205**, 999 (1965).
(2) Buser, W., Graf, P., Feitknecht, W., *Helv. Chim. Acta* **37**, 2322 (1954).
(3) Conway, B. E., "Electrochemical Data," Elsevier Publishing Co., Amsterdam, 1952.
(4) Davies, J. T., Rideal, E. K., "Interfacial Phenomena," p. 86, Academic Press, New York, 1961.
(5) Ghosh, B. N., Chakravarty, S. N., Kundu, M. L., *Chem. Soc.* **28**, 318 (1951).
(6) Grahame, D. C., *Chem. Rev.* **41**, 441 (1947).
(7) Healy, T. W., Herring, A. P., Fuerstenau, D. W., *J. Colloid Interface Sci.* **21**, 435 (1966).
(8) Koch, D. F. A., *Australian J. Chem.* **12**, 127 (1959).
(9) Kurbatov, M. H., Wood, G. B., Kurbatov, J. D., *J. Phys. Chem.* **55**, 1170 (1951).
(10) Morgan, J. J., Stumm, W., *J. Colloid Sci.* **19**, 347 (1964).

(11) Muller, J., Tye, F. L., Wood, L. L., "Batteries 2—4th International Symposium," p. 201, Pergamon Press, London, England, 1965.
(12) Wadsley, A. D., *J. Am. Chem. Soc.* **72,** 1781 (1950).
(13) Yopps, J. A., Fuerstenau, D. W., *J. Colloid Sci.* **19,** 61 (1964).
(14) Zabin, B. A., Taube, H., *Inorg. Chem.* **3,** 963 (1964).

RECEIVED October 26, 1967.

8

Adsorption of Selenite by Goethite

F. J. HINGSTON

Division of Soils, C.S.I.R.O., W. A., Laboratories, Wembley, Western Australia, 6014

A. M. POSNER and J. P. QUIRK

Department of Soil Science and Plant Nutrition, University of Western Australia, Nedlands, Western Australia, 6009

> *Specific adsorption of selenite on goethite increases the pH of the suspension and the negative charge on the oxide surface. Isotherms at constant pH are represented by,*
>
> $$\Gamma = \Gamma_{o(pH)} \cdot K_L(Se)/[1 + K_L(Se)],$$
>
> *where Γ = selenite adsorbed, $\Gamma_{o(pH)}$ = maximum adsorption, (Se) = equilibrium solution concentration of selenite, and K_L is a constant. $\Gamma_{o(pH)}$ and K_L vary with pH. The pH dependence of K_L is represented by,*
>
> $$K_L = K_2 K_D H^+/(H^+ + K_D),$$
>
> *where K_D = the second dissociation constant for selenious acid and $K_2 = K'_2/K_w$, where K'_2 is an exchange constant for OH^- and SeO_3^{2-} and K_w is the dissociation constant for water. The entropy gain of the reaction is consistent with release of a water molecule from the surface when a selenite ion is adsorbed.*

Adsorption of anions at mineral surfaces is important in soils because of the limit this process imposes on the availability of plant nutrients such as P, S, and Mo which occur naturally as anions and are added in anionic form in fertilizers. Anion adsorption is also relevant in geochemistry, ore processing, and other fields where minerals with high surface areas are brought into contact with aqueous solutions of anions. Selenite and goethite were chosen for this study because in Western Australia a selenium deficiency in pastures has been shown to be related to the incidence of white muscle disease in sheep (3), and according to workers quoted by Rosenfeld and Beath (9) selenium in soils of higher rainfall areas is probably present mainly as ferric selenites. Studies of selenite will supplement those of other anions such as phosphate and sulfate and should help to make it possible to formulate a general mechanism for adsorption of anions by soil colloids.

Methods

A micro-crystalline form of synthetic goethite consisting of aggregates of needle-like crystals with a B.E.T. surface area of 32 meter^2/gram was used as an adsorbent. The excess surface charge on this goethite was measured as a function of pH and ionic strength by the potentiometric method described by Parks and de Bruyn (8) except that NaCl was used as the supporting electrolyte in place of KNO_3. The amount of adsorption of selenite was measured radiometrically by counting tagged and standardized solutions. The effects of varying pH, supporting electrolyte concentration, reaction time and temperature were determined and the reversibility of the reaction was examined by isotopic exchange.

Results and Discussion

Reversibility. The reversibility of the adsorption reaction was tested by adding radioactive selenite to a suspension after reaction of the solid and inactive selenite. Adsorption was complete in one day, exchange of radioactive selenite with adsorbed selenite took seven days to come to equilibrium. Although exchange was slower than adsorption the same equilibrium value was reached, therefore the reaction is reversible under these conditions.

Adsorption Isotherms. The dependence of the amount of selenite adsorbed on pH and solution concentration of selenite is illustrated by the curves in Figures 1 and 2. These show that the amount of selenite taken up by goethite reaches a maximum value, $\Gamma_{o(pH)}$, at constant pH which cannot be exceeded by increasing the solution concentration and that this maximum value varies with pH. In the pH region studied ion size is unlikely to be the only factor limiting adsorption because even at low pH, where the maximum is greatest, the area of surface available to the ion is always greater than the area it would be expected to occupy ($\simeq 20$ A.^{2}/ion).

Isotherms calculated for constant pH (Figure 2) show that the amount of selenite adsorbed depends on the equilibrium solution concentration as represented by the equation of the Langmuir form,

$$\Gamma = \Gamma_{o(pH)} \cdot K_L(Se)/[1 + K_L(Se)] \qquad (1)$$

where $\Gamma_{o(pH)}$ is the maximum adsorption at the particular pH, (Se) is the concentration of selenite ions in solution, and K_L is the Langmuir constant.

The relationship between K_L and pH is shown in the graph of log K_L against pH (Figure 3), which is linear with a slope approaching unity at high pH and zero below pH 8—i.e., K_L becomes constant with decreasing pH.

Excess Surface Charge. The reaction at the goethite surface producing charged sites by adsorption of H^+ and OH^- as potential determining ions can be represented as follows,

$$\left[\begin{array}{c}\diagdown | \diagup OH_2 \\ Fe \\ \diagup | \diagdown OH_2\end{array}\right]^+ \underset{H^+}{\overset{OH^-}{\rightleftharpoons}} \left[\begin{array}{c}\diagdown | \diagup OH_2 \\ Fe \\ \diagup | \diagdown OH\end{array}\right]^0 \underset{H^+}{\overset{OH^-}{\rightleftharpoons}} \left[\begin{array}{c}\diagdown | \diagup OH \\ Fe \\ \diagup | \diagdown OH\end{array}\right]^-$$

Potentiometric titration of an aqueous suspension of oxides in the presence of varying concentrations of indifferent electrolyte has been used successfully to determine the zero point of charge (z.p.c.) and the variation in excess surface charge with pH (1, 8). The variation in excess surface charge (Γ_{H^+}-Γ_{OH^-}) with pH and NaCl concentration is shown for goethite in Figure 4.

The excess surface charge in the presence of specifically adsorbed ions was found by measuring the amount of selenite adsorbed and the amount of hydroxyl displaced into the solution. The quantity of OH^- displaced (Δ) was estimated for constant pH (a hypothetical situation) from the curves for titration of goethite, goethite plus selenite, and selenite alone by the equation.

$$(G + Se) - G\,Se = \Delta \qquad (2)$$

where G = titration value for goethite, Se = titration value for selenite, and GSe = titration value for goethite plus selenite, all expressed in μequiv./gram of goethite.

The excess surface charge can then be estimated from,

$$\delta Se = \delta - (Se^-)_{ads} + \Delta \qquad (3)$$

where δSe = charge in the presence of selenite, δ = charge in the absence of selenite, and $(Se^-)_{ads}$ = negative charge added by selenite ions allowing for the proportion of SeO_3^{2-} and $HSeO_3^-$ in solution. δSe was plotted against pH in Figure 4 to illustrate the decrease in z.p.c. and the variation in charge with ionic strength. Comparing δ with δ_{Se} it can be seen that adsorption of selenite always results in a decrease in the net charge—i.e., an increase in the net negative charge.

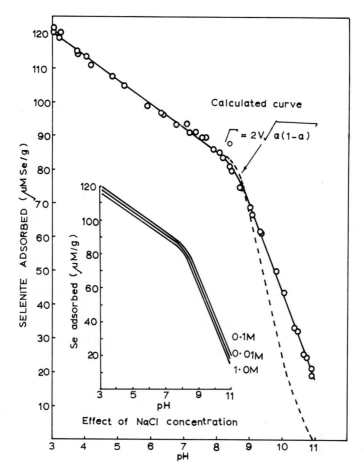

Figure 1. Relationship between pH and maximum amount of selenite adsorbed

Curve for ideal model shown thus - - - -, V = maximum adsorption of selenite at pK_D

A Model for the Mechanism of Adsorption. (1) ADSORPTION ENVELOPE. The curve relating $\Gamma_{o(pH)}$, the value for maximum adsorption at a particular pH, and pH is termed the "adsorption envelope."

Studies of specific adsorption of a series of anions to be discussed in detail elsewhere (5), have shown that the maximum specific adsorption at any pH, $\Gamma_{o(pH)}$, is related to the pK_D of the anion acid and the charge on the adsorbing species. From these studies essential requirements for specific adsorption of anions, or exchange of specifically adsorbed ions, seem to be as follows:

(a) A proton should be available either from a net excess on the surface or from a proton containing species in equilibrium with the anion in solution.

(b) An ionic species with a tendency to acquire a proton should be present (anions of weak acids have a tendency to acquire a proton at pH values near the pK_D of the acid).

(c) Specific adsorption of anions can only occur with an increase in the net negative charge on the surface.

Attempts to desorb selenite or any other specifically adsorbed anion (6, 7) by washing the solid with NaCl solutions of the same ionic strength and pH are frequently not successful. It has been found that leaching at constant pH restores the surface charge of an oxide to the value it had at that pH before specific adsorption occurred. In the case of selenite adsorbed on goethite, this occurs through desorption of OH^- rather than selenite. The selenite remaining when the charge has been restored can only be desorbed by increasing the negative charge through specific adsorption of another anion.

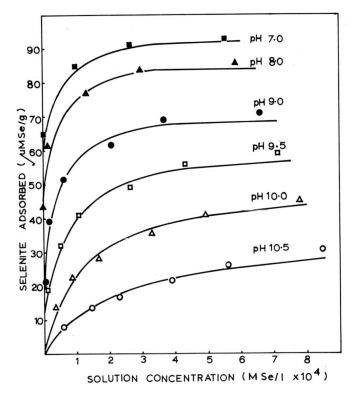

Figure 2. Langmuir isotherms

Experimental values shown by symbols and full lines calculated for best fit. pH values shown on curves

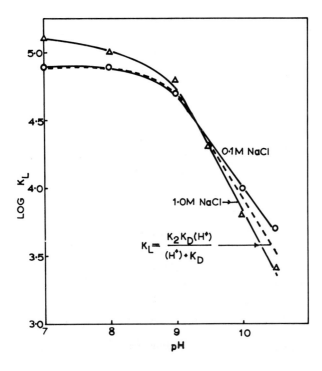

Figure 3. Plot of log K_L against pH: ○—0.1M NaCl, △—1.0M NaCl

---- Curve for ideal model 0.1M NaCl

At pH values higher than the z.p.c. maximum adsorption is determined by the proportion of the ions in solution that are able to donate and accept protons. The species SeO_3^{2-} can only accept a proton therefore it cannot be adsorbed in the absence of $HSeO_3^-$. For selenite in solution, the proportion of SeO_3^{2-} species is α, where α is the degree of dissociation of the species $HSeO_3^-$, and the proportion of $HSeO_3^-$ is $(1-\alpha)$. The probability of selecting SeO_3^{2-} from among the total selenite species is α and the probability of finding $HSeO_3^-$ is $(1-\alpha)$ therefore the probability of finding SeO_3^{2-} and $HSeO_3^-$ together is $\alpha(1-\alpha)$. Since this event can result in adsorption of both ions the amount of selenite adsorbed should be proportional to $\sqrt{\alpha(1-\alpha)}$ substituting for $\alpha = K_D/(H^+ + K_D)$, where K_D is the second dissociation constant for selenious acid, shows that $\Gamma_{o(pH)}$ is proportional to $\sqrt{K_D H^+/(H^+ + K_D)^2}$. The latter function reaches a maximum value at pH = pK_D. At more acid pH than pK_D the maximum adsorption is no longer dependent on the proportions of the species because $HSeO_3^-$ can both accept and donate protons. The maximum adsorption is then only limited by the

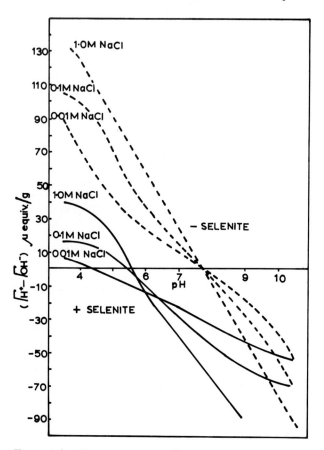

Figure 4. Excess surface charge and pH, selenite absent - - - -, selenite present ———, NaCl concentration indicated on curves

charge induced on the surface by adsorption. Deviation from this model could be expected to arise from interactions between ions on the surface and between ions and the surface. Interactions of this kind would probably be electrostatic and would result in regions of the envelope (Figure 1) having a Temkin form and the spread of the initial rise (high pH) over a greater range of pH than predicted by the simple model.

THE EXCHANGE ISOTHERM. For adsorption at constant pH the amount of selenite on the surface is related to the solution concentration by an equation of the Langmuir type (Equation 1).

If SeO_3^{2-} is the species determining adsorption and the reaction mechanism involves exchange with hydroxyl ions it can be shown that

$$K_L = K_2 K_D H^+ / (H^+ + K_D) \qquad (4)$$

where $K_2 = K'_2/K_w$, $K'_2 = $ the exchange constant for selenite and hydroxyl, and K_w = the dissociation constant for water. A curve using $pK_D = 9.2$ and the above relationship (Equation 4) is in good agreement with experiment (Figure 4). Although $pK = 9.2$ is somewhat higher than pK_2 for selenious acid in 0.1M NaCl—i.e., $pK_2 = 8.2$—it could be correct for the pK in the vicinity of the surface where an excess negative charge must be taken into account.

Substituting into Equation 1 gives,

$$\Gamma = \Gamma_{o(pH)} \cdot \frac{K_2 K_D H^+}{H^+ + K_D} (Se) \bigg/ \left[1 + \frac{K_2 K_D H^+}{H^+ + K_D} (Se) \right] \quad (5)$$

(3) APPROXIMATE THERMODYNAMIC QUANTITIES. Thermodynamic quantities, not taking into account activity coefficients, can be calculated from the experimental data at constant pH for the exchange reaction,

$$(OH^-)_s + SeO_3^{2-} \rightleftharpoons (SeO_3^{2-})_s + OH^- \quad (6)$$

The enthalpy of exchange $\Delta H_{1/2}$ referred to a standard state where the surface concentration $\Gamma = 1/2\ \Gamma_{o(pH)}$ was found using the van't Hoff equation on results from isotherms obtained at 5° and 20°C. Measurements on isotherms for pH 9 to pH 10.5 gave $\overline{\Delta H}_{1/2}$ of 4.5 ± 0.5 Kcal./mole. Values for the partial molar free energy of exchange, $\overline{\Delta G^*}_{1/2}$ (referred to standard state where $\Gamma = 1/2\ \Gamma_{o(pH)}$ and a hypothetical ideal molar solution of selenite and hydroxyl, less a configurational entropy term) can be calcuated from the value of K'_2 obtained from Equation 4. Since $K'_2 = 1.1 \pm 0.5$ and $\overline{\Delta G^*}_{1/2} = -RT \ln K'_2$, $\Delta G^*_{1/2} = 0.0_5 \pm 0.2_7$ Kcal./mole. An integral entropy $(\overline{\Delta S}_{1/2})$ of 16 ± 1 e.µ. was then obtained using the relationships,

$$\Delta S^*_{1/2} = \Delta H_{1/2} - \Delta G^*_{1/2}/T$$

and $\Delta S_{1/2} = \Delta S^*_{1/2} + 2 R \ln 2$ (4). If the reaction involves a change in the amount of water bound at the surface the equation giving entropy changes for the reactions is as follows,

$$S(SeO_3^{2-})_s + S(OH^-) + (x-y)S(H_2O)_s = S(SeO_3^{2-}) + S(OH)_s + \Delta S_{1/2}$$

where $S(i)$ = entropy of species "i" in solution and $S(i)_s$ = entropy of species "i" on the surface. Substituting values, $S(OH^-) = -10$ e.µ. and $S(SeO_3^{2-}) = -8$ e.u. given by Cobble (2) and the experimental value of $\Delta S_{1/2} = 16 \pm 1$ e.u. it follows that,

$$S(SeO_3^{2-})_s - S(OH^-)_s + (x-y)S(H_2O) = 18 \pm 1 \text{ e.u.}$$

Further, if the difference in entropy between selenite and hydroxyl is the same on the surface as it is in solution, the value for the entropy of water (16.7 e.u.) indicates that $(x - y) = 1$—i.e., a molecule of water is displaced from the surface during exchange of SeO_3^{2-} for OH^-. The value for K'_2 (1.1 ± 0.5) suggests that there is little difference in selectivity coefficients for selenite and hydroxyl on the goethite surface.

Literature Cited

(1) Atkinson, R. J., Posner, A. M., Quirk, J. P., *J. Phys. Chem.* **71**, 550 (1967).
(2) Cobble, J. W., *J. Chem. Phys.* **21**, 1443 (1953).
(3) Gardiner, M. R., *J. Dept. Agr. Western Australia*, 4th series, **4**, 632 (1963).
(4) Heath, N. S., Culver, R. V., *Trans. Faraday Soc.* **51**, 1575 (1955).
(5) Hingston, F. J., Atkinson, R. J., Posner, A. M., Quirk, J. P., *Nature* **215**, 1459 (1967).
(6) Kafkafi, U., Posner, A. M., Quirk, J. P., *Soil Sci. Soc. Amer. Proc.* **31**, 348 (1967).
(7) Muljadi, D., Posner, A. M., Quirk, J. P., *J. Soil Sci.* **17**, 212 (1966).
(8) Parks, C. A., de Bruyn, P. L., *J. Phys. Chem.* **66**, 967 (1962).
(9) Rosenfeld, I., Beath, O. A., "Selenium," Academic Press, New York, 1964.

RECEIVED October 26, 1967.

9

Coagulation by Al(III)

The Role of Adsorption of Hydrolyzed Aluminum in the Kinetics of Coagulation

HERMANN H. HAHN and WERNER STUMM

Division of Engineering and Applied Physics, Harvard University, Cambridge, Mass.

> *The kinetics of coagulation have been studied for systems of silica dispersions destabilized by hydrolyzed Al(III). The rate of agglomeration is a function of (1) the collision frequency which is determined by physical parameters such as colloid size and concentration and velocity gradients in the medium; and (2) of the collision efficiency factor which reflects the stability of the colloid. This relative stability has been determined as a function of chemical solution parameters such as pH and the ratio of coagulant concentration and surface concentration of the dispersed phase. The destabilization of silica dispersions results from specific adsorption of positively charged hydroxo aluminum complexes onto the negatively charged colloid surface causing a decrease and ultimately a reversal of sign of the surface potential.*

The aggregation of particles in a colloidal dispersion proceeds in two distinct reaction steps. Particle transport leads to collisions between suspended colloids, and particle destabilization causes permanent contact between particles upon collision. Consequently, the rate of agglomeration is the product of the collision frequency as determined by conditions of the transport and the collision efficiency factor, the fraction of collisions leading to permanent contact, which is determined by conditions of the destabilization step (2). Particle transport occurs either by Brownian motion (perikinetic) or because of velocity gradients in the suspending medium (orthokinetic). Transport is characterized by physical parameters. Particle destabilization is accomplished by different mechanisms of colloid chemical nature (Table I).

The purposes of this investigation were to gain a better insight into the means of destabilization by hydrolyzed Al(III) and to describe quantitatively the effect of the physical parameters and of the solution constituents on the rate of agglomeration. The influence of the pH, the amount of aluminum dosage, and the concentration of the colloidal phase upon particle agglomeration was evaluated in kinetic terms under controlled transport conditions (temperature, velocity gradient, number and dimension of colloid particles). The results indicate that silica dispersions are destabilized by specific adsorption of hydroxo aluminum complexes (19) onto the colloid surface. The physical parameters determine the particle transport and the collision frequency. The solution variables, influencing hydrolysis and specific adsorption of the destabilizing agent, affect the collision efficiency factor. The coagulation rate is given by the product of the collision frequency and the efficiency factor. The effect of chemical parameters on the transport step, owing to concentration changes which cause slight alterations in the viscosity of the suspending medium, are negligible. In a similar way, the influence of the temperature upon the rate of hydrolysis and polymerization of aluminum has no effect upon the collision efficiency as long as the rate of destabilization is much larger than the rate of transport.

Experimental

Colloids. Silica dispersions have been selected for this study because specific physical and surface chemical properties of the silica colloids are well defined and reproducible. SiO_2 particles are approximately spherical in shape and can be obtained in several sizes. The relative refractive index of amorphous silica particles with respect to various suspension media is described in the literature. The colloid chemical properties of silica in first approximation are similar to those of clays and of other solid material present in natural waters. The surface potential of silica colloids is negative within neutral pH ranges. The potential becomes increasingly negative when the pH is raised. Two types of silica were used: Ludox LS (E. I. DuPont de Nemours and Co.) with an average diameter of 15 mμ for experiments under perikinetic transport conditions and Min-U-Sil 30 (Pennsylvania Glass Sand Corp.) with an average diameter of 1.1 μ for the investigation of orthokinetic transport conditions. Particle diameters are determined from specific surface measurements.

Coagulant. Stock solutions were $10^{-2}M$ in reagent grade $Al(ClO_4)_3$ and $10^{-1}M$ in $HClO_4$. Aluminum determinations for the standardization of stock solutions were made by alkalimetric titration. The amount of residual dissolved aluminum in adsorption experiments was determined absorptiometrically using "aluminon" (14). The agreement between both

Table I. Schematic Representation of Different Steps in Particle Agglomeration (21)

Agglomeration of Colloidal Particles

Particle Agglomeration = Destabilization + Transport
Rate of Agglomeration = Coll. Efficiency Factor × Collision Frequency

COAGULATION

(1) *Reduction of potential energy of interaction between particles.*
(a) Compression of double layer by counterions (Schultze Hardy).
(b) Decrease in surface potential owing to specifically adsorbed counterions or owing to surface reactions.

FLOCCULATION

(2) *Bridge formation.* Specifically adsorbed polymeric species form bridges between colloids.

PERIKINETIC

(1) *Brownian motion* causes collision of primarily small colloids.

ORTHOKINETIC

(2) *Velocity gradients* transport large colloids, by imposing different velocities upon neighboring particles.

methods of aluminum determination was good for Al(III) concentrations larger than $10^{-4}M$. The coagulant was formed by raising slowly and uniformly the pH of the solution.

Ionic Medium. Silica dispersions were freshly prepared for each experiment in solutions buffered with $10^{-3}M$ HCO_3^-/CO_2. The amount of species dissolved from the amorphous silica surface during the experiment was negligible because of the small rate of dissolution reactions. The ionic medium in which coagulation and adsorption studies were carried out was kept constant: $I = 1.0$ to $2.0 \times 10^{-3}M$. The conditions in all agglomeration and adsorption experiments were such that no $Al(OH)_3$ precipitated within the period of observation.

Adsorption Experiments. The specific surface area of the dispersed phase was obtained from B.E.T. adsorption isotherms, using nitrogen as adsorbate. This method was described by Nelson and Eggertsen (11). The specific surface area found for Ludox LS is 205 meter^2/gram and for Min-U-Sil 30 2 meter^2/gram. The amounts of aluminum adsorbed onto the silica surface were determined from the difference between the total dosages of Al(III) added and the residual in the solution phase after a contact time of 5 minutes. The periods of contact were comparable to coagulation reaction times, and the results of the adsorption studies were expected to reflect the extent of adsorption occurring in the agglomeration process. Adsorption equilibrium was not necessarily attained.

Perikinetic Coagulation. The concentration of small silica colloids (Ludox LS $d = 15$ mμ) used in these studies was between 0.18 gram/liter and 0.43 gram/liter (36.5 — 86.0 meter^2/liter or 5.0-11.5 × 10^{13} particles/ml.). The dispersion, contained in a spectrophotometric cell, was mixed instantaneously with the destabilizing agent and allowed to coagulate. The change in the absorbance of light ($\lambda_o = 260$ mμ) resulting from changes in the particle size distribution of the coagulating dispersion is recorded as a function of time (Figure 1). One can determine the initial rate of coagulation, $-dN/dt$, from such recordings when the size of the dispersed colloids, d, is within the range given by Rayleigh's law: d $\leqslant 0.10 \lambda_o$ (Equation 4).

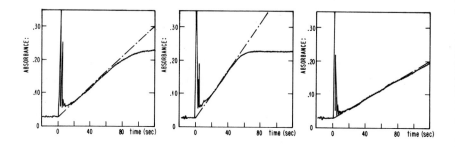

Figure 1. Measurement of perikinetic coagulation rate under various solution conditions

The rate of coagulation is obtained from a recording of the change in light absorbance ($\lambda_o = 260$mμ) with time. Light scattered by coagulating dispersions because of changes in particle agglomeration. Increase in light absorbance with increasing degree of agglomeration
(A) Ludox = 0.4 gram/liter; Al = 8.4 × 10^{-5}M; pH = 5.25; (B) Ludox = 0.4 gram/liter; Al = 8.4 × 10^{-5}M; pH = 6.6; (C) Ludox = 0.4 gram/liter; Al = 1.75 × 10^{-4}M; pH = 5.25

Orthokinetic Coagulation. Shear flow conditions, with average velocity gradients up to 100 sec.$^{-1}$, were generated reproducibly in a cylindrical reaction chamber with a turbine stirrer connected to a variable speed motor. The effective average velocity gradient or the actual collision frequency was obtained from a comparison of coagulation rates measured in this system and under well described perikinetic conditions. It was essential that the chemical conditions were the same in both experiments so that the collision efficiency factors were comparable. Dispersions were prepared from Min-U-Sil 30 (d = 1.1 μ) at concentration of 2 gram/liter (4 meter^2/liter or 10^9 particles/ml.). Samples were withdrawn in a microscopic counting cell and microphotographs were taken immediately after sampling. The decrease in total number of particles in suspension within the reaction time, $-dN/dt$, owing to agglomeration was determined by the counting of the total number of primary particles in each sample and comparing it with that fraction of primary particles incorporated into agglomerates (Figure 2). Other methods used for the evaluation of changes in the particle size distribution, such as light scattering techniques or Coulter Counter procedures, require the preparation of one or more dilutions prior to counting. Coagulant species may desorb partially as a result of dilution and such desorption would cause a change in the degree of agglomeration.

The Role of Adsorption in the Destabilization of Colloidal Dispersions by Hydrolyzed Al(III)

It is well known that hydrolyzed polyvalent metal ions are more efficient than unhydrolyzed ions in the destabilization of colloidal dispersions. Monomeric hydrolysis species undergo condensation reactions under certain conditions, which lead to the formation of multi- or polynuclear hydroxo complexes. These reactions take place especially in solutions that are oversaturated with respect to the solubility limit of the metal hydroxide. The observed multimeric hydroxo complexes or isopolycations are assumed to be soluble kinetic intermediates in the transition that oversaturated solutions undergo in the course of precipitation of hydrous metal oxides. Previous work by Matijević, Janauer, and Kerker (7); Fuerstenau, Somasundaran, and Fuerstenau (1); and O'Melia and Stumm (12) has shown that isopolycations adsorb at interfaces. Furthermore, it has been observed that species, adsorbed at the surface, destabilize colloidal suspensions at much lower concentrations than ions that are not specifically adsorbed. Ottewill and Watanabe (13) and Somasundaran, Healy, and Fuerstenau (16) have shown that the theory of the diffuse double layer explains the destabilization of dispersions by small concentrations of surfactant ions that have a charge opposite to

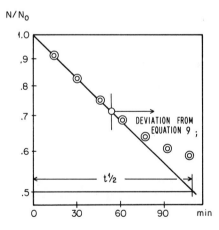

Figure 2. Evaluation of orthokinetic flocculation rate

The relative decrease in the total concentration of particles because of orthokinetic coagulation was determined from microscopic observation. The rate constant k_o is obtained from the slope of the semilogarithmic plot (Equation 9)
(Min-U-Sil 30 = 2 gram/liter; Al = 4.6 × 10^{-6}M; pH = 5.5; (du/dz) = 110 sec.$^{-1}$)

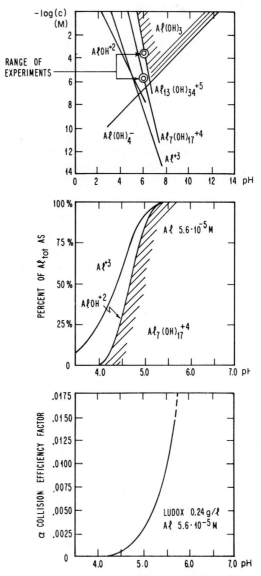

Figure 3. Hydrolysis of Al(III) and the effect of hydrolytic Al(III) species upon the coagulation rate

(A) Logarithmic diagram of Al(III) solubility as function of Al concentration and pH, derived from thermodyn. equilibrium constants. (B) Extent of Al-hydrolysis as function of pH. (C) Variation of the coagulation rate, expressed as collision efficieny factor, with pH at constant Al dosage (α values determined from Equation 3)

that of the colloids. However, one must include an energy component for specific adsorption in the total energy balance of the double layer interactions.

Hydrolysis of Al(III) and Destabilization Effects of Hydrolyzed Al(III). Figure 3 shows the observed correlation between the degree of hydrolysis of Al(III) and the effects of those aluminum species on the coagulation rate, expressed by a collision efficiency factor. Coagulation rates were measured under well defined perikinetic transport conditions and collision efficiency factors determined from a comparison with the known collision frequency under these conditions. Diagrams of the pH dependence of aluminum solubility and of the extent of aluminum hydrolysis under thermodynamic equilibrium conditions are presented together with a graph of the pH dependence of the measured collision efficiency factor, the fraction of all collisions that lead to permanent agglomeration. A comparison of these figures leads to the assumption that Al(III) becomes an efficient destabilizing agent when it is present in hydrolyzed and multimeric form. The diagrams show that pH and aluminum concentration were such that the solutions were oversaturated

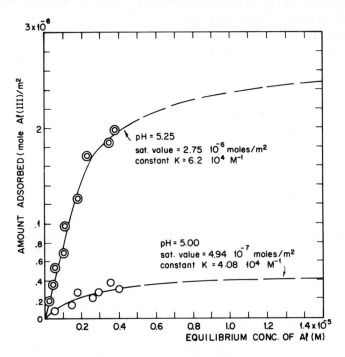

Figure 4. Adsorption of hydrolyzed Al(III) on colloidal silica

Curves are computed from Langmuir isotherms that have been fitted to the observed adsorption data

with respect to the solubility product of solid "Al(OH)$_3$." However, no precipitation of aluminum hydroxide was observed within the time span of the experiments. The species that cause effective destabilization of colloidal dispersions are therefore assumed to be soluble isopolycations. It must be pointed out in this connection that direct application of known hydrolysis equilibrium constants is not meanignful under conditions of oversaturation. Furthermore, the polynuclear species appearing in the diagrams of Figure 3 are only representative of the type of isopolycations that are possibly encountered under these circumstances. Neither thermodynamic equilibrium data nor information from adsorption and coagulation experiments permits at the present a more quantitative description of these multinuclear hydroxo complexes.

The Adsorption of Hydrolyzed Al(III). O'Melia and Stumm (*12*) have shown that specific adsorption of hydrolyzed Fe(III) species accounts for the observed coagulation and restabilization of silica dispersions. A model was formulated on the basis of the Langmuir adsorption isotherm and was shown to explain the observations adequately. The authors derive a relationship between the surface area concentration of the dispersed phase S (in meter^2/liter and the applied coagulant ion concentration $C_{t\theta}$ (in M) necessary to reach a certain fraction of surface coverage. The extent of destabilization or of restabilization can be concluded from the amount of surface coverage on the colloidal particle:

$$C_{t\theta} = \frac{\theta}{K(1-\theta)} [1 + KS\Gamma_m(1-\theta)] \tag{1}$$

Γ_m = sorption capacity (moles of metal ion/meter^2)
K = Langmuir's adsorption-desorption equilibrium constant (M^{-1})
θ = fraction of surface covered—ratio of the amount of metal ion adsorbed at colloid surface and the total sorption capacity (dimensionless)

The adsorption of hydrolyzed Al(III) on silica is described by the isotherms in Figure 4. The data can be fitted by the Langmuir equation as was the case for the adsorption of hydrolyzed Fe(III) on silica (*12*). This agreement between experimental data and the Langmuir isotherm, derived for an adsorption-desorption equilibrium, cannot be used to conclude that adsorption in this instance is fully reversible or that equilibrium had been reached in the experiments. Figure 5 compares the adsorption of hydrolyzed Al(III) onto silica and the effects of addition of multimeric Al(III) hydroxo complexes to colloidal dispersions on the coagulation rate of these suspensions, as expressed by a change in the collision efficiency factor. The following qualitative explanations for the observed destabilization and restabilization of colloidal silica with hydrolyzed Al(III) can be derived from Figure 5: (1) The colloidal suspension becomes unstable, as shown by collision efficiency factors larger than zero and begins to coagulate at the critical coagulation concentration (c.c.c.), where a certain minimum surface coverage on the colloid occurs. The experimental evidence supports the assumption, that specific adsorption of isopolycations on the colloid surface leads to a decrease of the surface potential to a critical threshold value and causes destabilization; (2) further increase in the coagulant concentration C_t brings about a restabilization of the dispersion, shown by decreasing collision efficiency factors (extrapolation to zero efficiency factor gives the critical stabiliza-

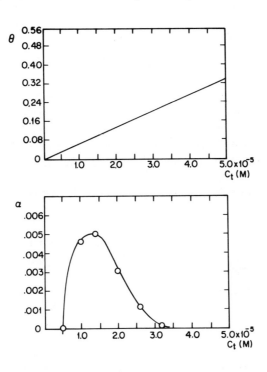

Figure 5. *The variation of relative colloid stability, expressed as collision efficiency factor, with Al(III) dosage as compared with colloid surface coverage from adsorption of Al(III)*

(A) Surface coverage as a function of total amount of Al(III) added, C_t, at constant pH—computed on the basis of a Langmuir adsorption isotherm (K = 6.2 × $10^{-4}M^{-1}$, Γ_m = 2.75 × 10^{-6} moles/meter^2, Ludox LS = 0.3 gram/liter

(B) Relative colloid stability as a function of the total concentration of coagulant added, C_t, at constant pH—obtained from the measurement of perikinetic coagulation rates according to Equation 3. (Ludox LS = 0.3 gram/liter)

tion concentration (c.s.c.). Large amounts of specifically adsorbed cationic hydroxo aluminum complexes cover the originally negatively charged surface to a large extent and cause reversal of surface charge and potential. The colloidal particles repel each other electrostatically under these conditions and suspensions are stable again. Figure 5 shows that the extent of surface coverage is very small at the point of beginning sol destabilization ($\theta \cong 0.03$ at pH = 5.25) and that the collision efficiency factor, characterizing the instability of the colloids, decreases again for surface coverage values larger than $\theta \cong 0.10$ (pH = 5.25). These observations lead to the conclusion that bridging between colloidal particles by means of polymeric hydrolysis products does not play a significant role in the destabilization with hydrolyzed Al(III) at these pH values.

Figure 3 indicates that the character of the hydrolysis species is strongly dependent upon the pH of the solution. This may explain the large effects of the pH of the solution upon the sorption tendency of the hydrolyzed aluminum species. Figure 6 summarizes the adsorption data and shows that the total sorption capacity Γ_m increases markedly when the pH is raised in the range from 4.5 to 5.25. Coagulation experiments have shown that the amount of hydrolyzed Al(III) necessary to obtain collision efficiency factors larger than zero (c.c.c.), and the dosage at which maximum collision efficiency factors occur ($C_{t\,opt}$) are constant within the pH range from 4.5 to 5.25 (c.c.c. = $4 \times 10^{-7} \pm 10^{-7}$ moles Al(III)/meter^2 SiO$_2$ and $C_{t\,opt} = 10^{-6} \pm 10^{-7}$ moles Al(III)/meter^2 SiO$_2$). The increase with pH in the sorption capacity and the constancy of the necessary coagulant dosages over a certain pH range explain the decrease in the fractional surface coverage necessary to reduce the surface potential of the colloids to a critical threshold value or to effect a total reversal of the potential.

The approximate constancy of c.c.c. and $C_{t\,opt}$ within the investigated pH range is in accord with the hypothesis that the destabilization of colloidal suspensions is caused by specific adsorption of hydrolyzed Al(III) on the colloid surface. The observation suggests that the dosage of Al(III) necessary to effect a certain reduction of the surface potential of the colloid is relatively independent of the nature of the hydrolysis species. This can be explained in part by the fact that the ligand number of aluminum hydroxo species (OH$^-$ bound per Al(III)-ion), which is indicative of an average net charge on the ion, remains nearly unaffected by changes in the solution pH within certain limits (18). It must be concluded from all reported observations that the destabilization by hydrolyzed Al(III) is presumably the result of a partial or complete neutralization of the negative surface charge by positively charged multinuclear aluminum hydroxo complexes specifically adsorbed on the silica surface. Results of a recent investigation on the electrophoretic mobility of silica colloids at different C_t and pH, by Kane, LaMer, and Linford (5) support this model of adsorption coagulation.

Forces of Adsorption. The mechanism of attachment of isopolycations on solid surfaces and the nature of the forces that hold the adsorbate are not well understood. However, Stumm and O'Melia (20) have given a few qualitative arguments that explain in part the observed specific adsorption of hydrolyzed Al(III) on silica. First, the effective charge density of the central ion is reduced with increasing degree of hydroxidation and multimerization. Hydrolyzed species are larger and to a smaller extent hydrated than non-hydrolyzing ions. The hydroxo aluminum complexes are less "hydrophilic" if compared with the strongly hydrated, non-hydrolyzed aluminum ion. Correspondingly, these isopolycations will accumulate at the solid solution interface. Results from studies on the role of hydration in the adsorption of certain earth alkali ions onto quartz by Malati and Estefan (6) supplement these qualitative arguments. Second, these multimeric species contain more than one OH^- group per ion that can become attached at the colloid surface. Iso- and heteropolyanions (polysilicates, silicotungstates) have been found to show a similar tendency for specific adsorption as isopolycations.

Kinetics of Coagulation: Particle Transport as Rate Determining Step

It is known that the rate of coagulation can be increased by increasing the concentration of the suspended solid mass or by stirring the suspension more intensively. On the other hand, solution variables have seemingly a pronounced effect on the rate of particle agglomeration, as indicated in Figures 3 and 5. These observations can be explained, when the different reaction steps in the coagulation process and the relative effect of these reactions on the observed coagulation rate, $-dN/dt$, are considered.

One can distinguish the following steps in the agglomeration of silica dispersions with hydrolyzed aluminum: (1) hydrolysis and multimerization of Al(III) to isopolycations; (2) diffusion of these aluminum hydroxo complexes to the colloid surface and adsorption; and (3) transport of suspended particles and collision resulting in certain instances in attachment of colloids. The transport of colloids to each other has been observed to proceed more slowly than all other steps under the given circumstances (3). It follows that the rate of agglomeration, $-dN/dt$, is obtained from the collision frequency, as determined solely by the transport mechanism, corrected by a collision efficiency factor, that describes the fraction of collisions resulting in permanent agglomeration. The value of the collision efficiency factor depends upon the extent of destabilization of the dispersion.

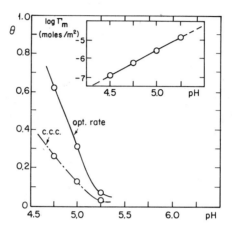

Figure 6. *Sorption capacity and surface coverage by hydrolyzed Al(III) on colloidal silica as function of pH*

The values given have been computed on the basis of Langmuir isotherms that were fitted to the experimentally observed data. (Ludox LS = 0.3 gram/liter)

Perikinetic Coagulation. If colloidal particles are of such dimensions that they are subject to thermal motion, the transport of these particles is accomplished by this Brownian motion. Collisions occur when one particle enters the sphere of influence of another particle. The coagulation rate measuring the decrease in the concentration of particles with time, N (in numbers/ml.), of a nearly monodisperse suspension corresponds under these conditions to the rate law for a second order reaction (15):

$$-dN/dt = k_p N^2 \qquad (2)$$

The rate constant k_p is given in terms of physical parameters (Boltzmann Constant K_B, the absolute temperature T, and the absolute viscosity η) that characterize these transport conditions. In the case of not completely destabilized colloids, when according to v. Smoluchowski so-called slow coagulation is observed, the rate constant contains in addition the collision efficiency factor, α_p, the fraction of collisions leading to permanent attachment under perikinetic conditions:

$$k_p = \alpha_p \frac{4K_BT}{3\eta} \tag{3}$$

According to this kinetic model the collision efficiency factor α_p can be evaluated from experimentally determined coagulation rate constants (Equation 2) when the transport parameters, K_BT, η are known (Equation 3). It has been shown recently that more complex rate laws, similarly corresponding to second order reactions, can be derived for the coagulation rate of polydisperse suspensions. When used to describe only the effects in the total number of particles of a heterodisperse suspension, Equations 2 and 3 are valid approximations (*4*).

Changes in the cumulative particle size distribution, describing the decrease in the total number of dispersed colloids caused by agglomeration have been determined in this instance by a light scattering method. Troelstra and Kruyt (*23*) have derived an equation for the light absorbance, A, describing scattering effects of coagulating colloids, on the basis of Rayleigh's law:

$$\text{Abs} = BN_oV_o^2(1 + 2k_pN_ot) \tag{4}$$

B = optical constant of scattering system, depending upon the refractive indices of the solid scatterers and the suspending medium, the wavelength of the light used and the length of the light path;

N_o, V_o are the concentration and volume of the colloidal particles at reaction time $t = 0$;

k_p = rate constant of second order reaction, defined in Equation 3;

The equation derived by Troelstra and Kruyt is only valid for coagulating dispersions of colloids smaller than a certain maximum diameter given by the Rayleigh condition, $d \leq 0.10 \lambda_o$. Equation 4 applies in cases where particles are transported solely by Brownian motion. Furthermore, the kinetic model (Equations 2 and 3) has been derived under the assumption that the collision efficiency factor does not change with time. In the case of some partially destabilized dispersions one observes a decrease in the collision efficiency factor with time which presumably results from the increase of a certain energy barrier as the size of the agglomerates becomes larger.

All the assumptions discussed hold during the first part of the reaction. Figure 1, a recording of the change in light absorbance with time of a coagulating dispersion, shows a straight line relationship according to Equation 4 for short reaction periods. Only after extended reaction times does a deviation from predicted values occur. Consequently, only the initial coagulation rate and the incipient collision efficiency factor can be evaluated from these measurements by determining the quotient of slope over ordinate intercept of the initially straight line. The total recorded reaction times in these experiments are in the order of 3 to 5 minutes.

Orthokinetic Coagulation. If the colloidal particles are very large in comparison with molecules, the transport of the colloids to each other will depend predominantly upon velocity gradients within the medium of the suspension, resulting from agitation or convective currents. The effects of Brownian motion on these dispersions are comparatively small. Von Smoluchowski (15) has given an equation for the number of collisions between colloids of a monodisperse suspension subject to velocity gradients, resulting from shear flow of viscous media. The rate of coagulation, expressed as a decrease of the total number of particles in suspension per unit time is found from the collision frequency, as determined by the transport conditions and the collision efficiency factor, which depends upon the extent of destabilization of the suspension. The resulting rate law for nearly monodisperse suspensions corresponds to a first order reaction:

$$-dN/dt = k_o N \tag{5}$$

The rate constant k_o for orthokinetic coagulation is determined by physical parameters (velocity gradient du/dz, floc volume ratio of the dispersed phase, ϕ = sum over the product of particle number and volume), and the collision efficiency factor α_o observed under orthokinetic transport conditions:

$$k_o = \alpha_o \frac{4(du/dz)\phi}{\pi} \tag{3}$$

The overall rate of agglomeration of any suspension, that is prepared experimentally and consists of small and large colloids, is obtained by adding the expressions derived for perikinetic and orthokinetic coagulation (Equations 2 and 5):

$$-dN/dt = \alpha_p \frac{4K_B T}{3\eta} N^2 + \alpha_o \frac{4(du/dz)\phi}{\pi} N \tag{7}$$

From a comparison of the two collision frequency terms, described in detail in the Equations 3 and 6, one obtains the relative contributions of the perikinetic and orthokinetic transport to the total particle agglomeration. The ratio is a function of the radius of the colloid, r, and the absolute value of the velocity gradient du/dz:

$$\frac{k_{\text{orthokinetic}}}{k_{\text{perikinetic}}} = \frac{4r^3 \eta (du/dz)}{K_B T} \tag{8}$$

It can be shown with Equation 8 that both transport mechanisms contribute equally to the coagulation of a Ludox LS dispersion, containing silica colloids of an average diameter of 15 mμ, only when the velocity gradients are extremely large, in the order of 3 $\times$ 10^5 sec.$^{-1}$. This leads to the conclusion that agglomeration of sufficiently destabilized Ludox LS dispersions occurs in its early stages through Brownian motion alone. The physical variables in Equation 3 therefore are unambiguously defined and permit a determination of the collision efficiency factor from the measured rates. Similarly, at average velocity gradients of about 100 sec.$^{-1}$ which were generated in the shear flow apparatus, all particles larger than 215 mμ are predominantly transported by velocity gradients. Again it can be concluded that the coagulation of Min-U-Sil 30 dispersions with colloids of diameter 1.1 μ follows the orthokinetic rate law and the collision efficiency value can be determined according to Equation 6.

The actual decrease of the total number of particles of a dispersion coagulating under orthokinetic conditions is described by a logarithmic function obtained from the integration of Equation 5:

$$\ln(N/N_o) = -\alpha_0 \frac{4(du/dz)\phi}{\pi} t \tag{9}$$

Figure 2 is a semilogarithmic graph of the microscopically determined decrease with time of the total number of particles, relative to the initial number of colloids, caused by orthokinetic coagulation. It depicts the expected logarithmic decrease for short reaction times and a deviation from values predicted by the kinetic model after extended reaction periods. Presumably, this results from the increasing polydispersity of the suspension and the decreasing collision efficiency factor with time. Initial coagulation rates and collision efficiency factors are obtained from the slope of the incipiently straight line. Reaction times were in these instances in the order of 1 to 2 hours.

The collision efficiency factors, describing the extent of the colloid destabilization, within certain limits are equal under perikinetic and orthokinetic conditions (3).

The Effect of Solution Variables on the Coagulation Rate

Chemical parameters determine the surface characteristics of the suspended colloids, the concentration of the coagulant and its effects upon the surface properties of the destabilized particles, and the influence of other constituents of the ionic medium upon the coagulant and the colloids. The extent of the chemical and physical interactions between the colloidal phase and the solution phase determines the relative stability of the suspended colloids. One speaks of stable suspensions when all collisions between the colloids induced by Brownian motion or by velocity gradients are completely elastic: the colloidal particles continue their separate movements after the collision. A dispersion is theoretically completely destabilized, when all collisions between the suspended particles lead to permanent attachment. Thus, the collision efficiency factor is a quantitative measure of the relative stability of a sol. Together with the total number of collisions occurring between the particles, the collision frequency, which is determined by the temperature, the velocity gradient and the number concentration and dimension of the colloidal particles, the collision efficiency factor describes the observed coagulation rate (Equation 7).

Colloid Stability as a Function of pH, C_t, and S. The effects of pertinent solution variables (pH, Al(III) dosage C_t, Al(III) dosage relative to surface area concentration of the dispersed phase S upon the collision efficiency, have been determined experimentally for silica dispersions and hydrolyzed Al(III). However, one cannot draw any conclusion from the experimental results with respect to the direct relationship between conditions in the solution phase and those on the colloid surface. It has been indicated by Sommerauer, Sussman, and Stumm (*17*) that large concentration gradients may exist at the solid solution interface which could lead to reactions that are not predictable from known solution parameters.

Figure 7 gives the relative stability of SiO_2 colloids in terms of the observed collision efficiency as a function of Al(III) dosage C_t for different pH values. Such data can be used to define quantitatively similar log C_t-pH domains of coagulation and restabilization, expressed as the critical concentrations of beginning destabilization and complete restabilization (c.c.c. and c.s.c.) as given by Matijević *et al.* (*8, 9, 10*). A similar procedure for the determination of the c.c.c. and c.s.c. concentration limits has been proposed by Teot (*22*).

The variation of α with C_t at constant pH can be explained by the change in destabilizing effects of different concentrations of adsorbed coagulant at the colloid surface. α initially increases when C_t becomes larger because of increasing amounts of specifically adsorbed isopolycations C_a, that cause a progressive decrease of the surface potential of the colloid. Further increase in C_t and consequently C_a will eventually cause charge neutralization, indicated by maximum α. The surface potential becomes now more and more negative because of continued specific adsorption of isopolycations. This explains the decrease in α for large coagulant concentrations. The non-linear relationship between C_t and C_a, especially when the sorption capacity is already nearly exhausted, may primarily account for the asymmetrical shape of the function describing the variation of α with C_t. It is possible that other factors, such as changes in the electrostatic repulsion of the SiO_2 colloids or their

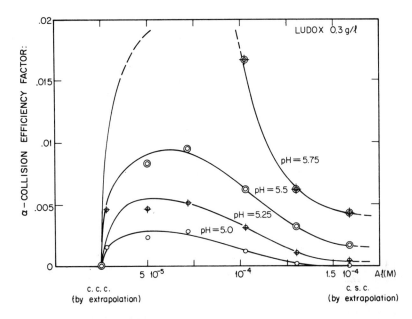

Figure 7. Effects of pH and Al(III) dosage upon relative colloid stability, expressed as collision efficiency factor, for constant surface concentration

Diagrams such as these can be used to establish pH, Al(III),S domains, describing the limits between stable, unstable, and restabilized suspensions

originally strongly hydrophilic character when the colloid surface is covered to a large extent with isopolycations, also contribute to the observed difference in dosages necessary to obtain a certain degree of destabilization and restabilization. Teot (22) using Al(III) as coagulant and O'Melia and Stumm (12) studying coagulation with Fe(III) also find that the optimum coagulant dosage $C_{t\,opt}$ resulting in maximum collision efficiency is much closer to c.c.c. than to c.s.c. The increase of α with pH at the point of optimum coagulant dosage, where reversal in the sign of the surface potential occurs, is not too well understood. A qualitative explanation is perhaps that the decrease in surface coverage at $C_{t\,opt}$, as the pH increases, permits particles to approach each other more closely upon collision. This increases the effect of the London-van-der-Waals forces of attraction on the colloids. The change in the nature of the hydroxo aluminum complexes with pH may be another factor accounting for the decrease in colloid stability at $C_{t\,opt}$.

Figure 8 illustrates the relationship between the surface concentration of the dispersed phase S and the coagulant dosage C_t necessary to affect a certain degree of colloid instability at constant pH. If the described model of destabilization is correct, one should observe a direct relationship between S and C—*i.e.*, the larger the surface concentration of the dispersed phase the more coagulant has to be added to attain certain α-values. The so-called stoichiometric relationship between S and C_t, which can be formulated on the basis of a Langmuir adsorption isotherm (12), has been found in these experiments.

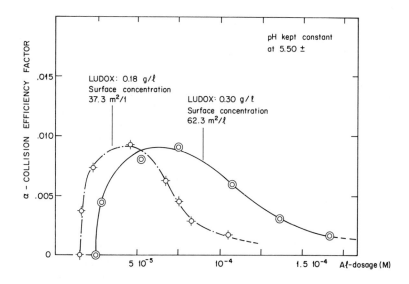

Figure 8. Stoichiometric relationship between surface concentration of colloidal phase and necessary Al(III) dosages

An increase in colloid concentration requires increased Al(III) dosages to obtain comparable values of relative colloid stability

Concluding Remarks

The rate of coagulation depends upon the collision frequency, which is controlled by physical parameters describing perikinetic or orthokinetic particle transport (temperature, velocity gradient, number concentration and dimension of colloidal particles), and the collision efficiency factor α measuring the extent of the particle destabilization which is primarily controlled by chemical parameters.

The colloidal silica dispersions are destabilized with hydrolyzed Al(III) primarily because of adsorption of polyhydroxo aluminum cations on the colloid surface which reduces the incipiently negative surface potential of SiO_2 colloids (adsorption coagulation). The pertinent solution variables describing the destabilization reaction are pH, total aluminum concentration C_t, and the ratio of aluminum dosage to the colloid surface concentration S.

The rate of coagulation by Al(III) can be improved operationally: (1) by increasing the collision frequency through raising the velocity gradient and (2) by adjusting the solution variables (pH, C_t, S) such that the collision efficiency factor becomes optimal. Figure 9 schematically illustrates how physical and chemical factors affect the coagulation rate.

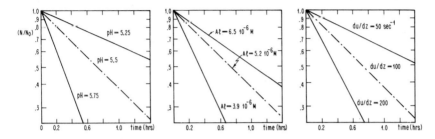

Figure 9. *Relative effects of collision efficiency factor and velocity gradient on coagulation rate*

The coagulation rate depends upon physical parameters (temperature, velocity gradient, number and dimension of colloid), determining the collision frequency and upon chemical parameters (pH, Al(III) dosage, surface concentration of dispersed phase S), affecting the collision efficiency factor α
(A) $Al = 5.2 \times 10^{-6} M$, $(du/dz) = 100$ sec.$^{-1}$; (B) $pH = 5.5$, $(du/dz) = 100$ sec.$^{-1}$; (C) $pH = 5.5$, $Al = 5.2 \times 10^{-6} M$; MIN-U-SIL 30 = 1 gram/liter

It must be kept in mind that the efficiency of the coagulation process in practice is not solely determined by the agglomeration rate; the attainment of certain desirable floc properties must be included in deliberations directed toward the optimization of the process.

Nomenclature

A	= Light absorbance in 4 cm. cell
α	= Collision efficiency factor
α_o	= Collision efficiency factor, measured under orthokinetic conditions
α_p	= Collision efficiency factor, measured under perikinetic conditions
B	= Optical constant of scattering system (Rayleigh constant)
C_a	= Concentration of adsorbed coagulant [M]
C_t	= Total concentration of coagulant added [M]
$C_{t\theta}$	= Total concentration of coagulant necessary to attain surface coverage θ [M]
$C_{t\,opt}$	= Total concentration of coagulant necessary to attain optimum α [M]
c.c.c.	= Critical coagulation concentration [M]
c.s.c.	= Critical stabilization concentration [M]
d	= Diameter of colloidal particle [μ] or [mμ]

Γ_{max}	= Amount of coagulant adsorbed at surface saturation [M/meter^2]
I	= Ionic strength of suspending medium [M]
K	= Langmuir's adsorption equilibrium constant [M^{-1}]
K_B	= Boltzmann constant
k_o	= Reaction rate constant of orthokin. coagulation [sec.$^{-1}$]
k_p	= Reaction rate constant of perikin. coagulation [sec.$^{-1}$ cm.^{3}]
λ_o	= Wavelength of light used in scattering experiments [mμ] (in vacuum)
N	= Total concentration of particles suspended [numbers per ml.]
$N_o V_o$	= Total concentration of particles suspended; volume of particles suspended; at time $t = 0$
dN/dt	= Coagulation rate [sec.$^{-1}$ cm.^{3}]
η	= Absolute viscosity of suspending medium
S	= Surface area concentration of colloidal phase [meter^2/liter]
T	= Absolute temperature
θ	= Fractional surface coverage
ϕ	= Floc volume ratio of dispersed phase

Acknowledgment

This work was supported in part by Research Grant WP 000 98 of the U.S. Public Health Service and the Federal Water Pollution Control Administration.

Literature Cited

(1) Fuerstenau, M. C., Somasundran, P., Fuerstenau, D. W., *Trans. Inst. Mining Met.* **74**, 381 (1965).
(2) Hahn, H. H., *Jahrbuch vom Wasser* **33**, 172 (1966).
(3) Hahn, H. H., Stumm, W., *J. Colloid Interface Sci.* **28**/1, 134 (1968).
(4) Hidy, G. M., *J. Colloid Sci.* **20**, 123 (1965).
(5) Kane, J. C., LaMer, V. K., Linford, H. B., *J. Colloid Interface Sci.* (in print).
(6) Malati, M. A., Estefan, S. F., *J. Colloid Interface Sci.* **22**, 306 (1966).
(7) Matijević, E., Janauer, G. E., Kerker, M., *J. Colloid Sci.* **19**, 333 (1964).
(8) Matijević, E., et al., *J. Phys. Chem.* **57**, 951 (1953).
(9) *Ibid.*, **64**, 1175 (1960).
(10) *Ibid.*, **65**, 826 (1961).
(11) Nelson, F. M., Eggertsen, F. T., *Anal. Chem.* **30**, 1387 (1958).
(12) O'Melia, C. R., Stumm, W. J., *J. Colloid Interface Sci.* **23**, 437 (1967).
(13) Ottewill, R. H., Watanabe, A., *Kolloid Z.* **170**, 38 (1959).
(14) Packham, R. F., *Proc. Soc. Water Treat. Exam.* **7**, 102 (1958).
(15) v. Smoluchowski, M., *Z. Phys. Chem. (Leipzig)* **92**, 129 (1917).
(16) Somasundaran, P., Healy, T. W., Fuerstenau, D. W., *J. Colloid Interface Sci.* **22**, 599 (1966).
(17) Sommerauer, A., Sussman, D. L., Stumm, W., *Kolloid Z. und Z. Polymere* (in print).
(18) Stumm, W., Morgan, J. J., *JAWWA* **54**, 97 (1962).

(19) Stumm, W., "Principles and Applications of Water Chemistry," p. 520, S. D. Faust, J. V. Hunter, Eds., John Wiley & Sons Inc., New York, 1967.
(20) Stumm, W., O'Melia, C. R., *JAWWA* **60,** 514 (1968).
(21) Stumm, W., Hahn, H. H., *Symp. Proc. Coagulation Flocculation, Univ. of Karlsruhe, Germany* (September 1967).
(22) Teot, A. S., *Conf. Polymer Sci., New York Acad. Sci., New York* (May 1967).
(23) Troelstra, S. A., Kruyt, H. R., *Kolloidchem. Beihefte* **54,** 225 (1943).

RECEIVED November 24, 1967.

10

Reaction of the Hydrated Proton with Active Carbon

VERNON L. SNOEYINK and WALTER J. WEBER, JR.

Department of Civil Engineering, College of Engineering, The University of Michigan, Ann Arbor, Mich.

> *Reactions of the hydronium ion with porous active carbon have been investigated in aqueous systems. Hydronium-ion activity, specific-anion concentration, and carbon dosage have been among the major variables studied. Rates of reaction have been found to be limited by pore diffusion, as partially verified by activation energies of −(2 to 3) kcal. per mole-deg. The results can be interpreted partly in terms of a reaction of the hydronium ion and dissolved oxygen with a surface benzpyran (chromene) group to produce hydrogen peroxide and a surface benzopyrylium (carbonium) ion with a sorbed anion, and partly in terms of physical sorption of the acid on the carbon surface.*

Hydronium ion concentration has been found to be a significant factor in the adsorption of various organic compounds from aqueous solution by active carbon (28). A partial explanation of this effect is afforded by the fact that the ionization—and therefore the mobility and adsorptive properties—of many organic molecules is affected by pH. However, pH changes have been found to affect the uptake of certain organic molecules under circumstances in which these changes would not have a significant effect on the ionic character of the adsorbing species. Weber and Morris (28) have found increased rates of adsorption with decreasing pH for sulfonated alkylbenzenes in pH regions far removed from the pK range for these compounds. The enhanced adsorption rates have been attributed to partial neutralization of the active carbon's negative surface charge, thus reducing resistance to pore transport. Studies on rates of uptake of various substituted phenols also have indicated that the pH effect is more than can be explained in terms of simple variations in sorbate species (11).

Several studies have been directed toward examination of the interaction of acids and bases with active carbons (*1, 8, 10, 17, 18, 19*). Boehm (*3*), Garten and Weiss (*9*), and Snoeyink and Weber (*21*) have presented reviews on the subject. Garten and Weiss (*8, 9, 10*) have shown that acid and alkali sorption can be related to surface functional groups which form during the preparation of the carbon. Alkali sorption occurs principally on carbons activated at temperatures near 400°C., and is attributed to the presence of phenolic and lactone functional groups on the carbon surface. Carbons which sorb acid usually are activated at temperatures near 1000°C.; the acid reaction in this case is assumed to take place with chromene (benzpyran) structures on the surface.

The studies reported in this paper have focused on more complete elucidation of the nature of the interaction between the hydronium ion and active carbon. Both rate and extent of reaction have been studied as a function of several variables to obtain data which ultimately should contribute to a meaningful interpretation of pH effects on adsorption of organic solutes by active carbon.

Experimental

Carbons. The active carbon used for the majority of the experiments in this study was a granular, commercial coconut-shell carbon, carefully sieved to a size range which included discrete particles passing a U.S. Standard Sieve No. 50 and being retained on a No. 60 sieve; the mean particle diameter for this size range is 273 microns. After sieving, the carbon was washed thoroughly with distilled water to remove dust and fines, and then dried to a constant weight at 105°C. The inorganic content was 0.7% by weight. One of the primary reasons for choosing this particular carbon was its resistance to attrition in the rapidly-stirred experimental reactors.

A coal-base carbon, prepared in the same fashion and of the same mean particle size, was employed in several experiments. The maximum ash content for this material, as reported by the manufacturer, was 8% (*20*).

A pore size distribution was not available for the coconut-shell carbon, but, again according to the manufacturer, approximately 55% of the intraparticle volume of the coal-base carbon was comprised of diameters between 15 and 20 A. (*20*). The coal-base carbon was designed primarily for adsorption from solution, while the coconut-shell carbon was intended primarily for application in gaseous systems (*6*).

Experimental Systems. Rate-of-reaction studies were carried out utilizing both finite and infinite bath techniques. Test solutions were prepared at the desired ionic strength, temperature, and initial pH. These solutions were stirred rapidly with a motor-driven polyethylene-coated stirring blade. For each test, a carefully measured quantity of carbon was added in the dry form. The finite bath technique consisted of recording pH values as a function of time after carbon addition. All finite bath pH measurements were made with a Corning Model 12, expanded-scale pH meter. The infinite bath technique consisted of maintaining a constant pH throughout the reaction with a Sargent recording pH Stat. As the reaction proceeded, the pH Stat continuously added and recorded the quantity of standard acid ($0.1N$ in this case) needed to maintain a constant pH. Stirring speeds in all cases were in excess of the minimum required to keep the carbon in suspension; separate experiments indicated independence of sorption rate on stirring speeds greater than this minimum.

Equilibrium studies were performed to determine the extent of the hydronium ion-active carbon reaction. These studies were carried out in reaction vessels containing 1.5 liters of distilled water adjusted to the desired pH and ionic strength. Temperature control was provided by immersing the reactor in a water bath, except for tests at room temperature ($25° \pm 3°C.$). At the start of each experiment a known quantity of carbon was added to the test solution. Stirring was again accomplished with a motor-driven polyethylene stirring blade at speed sufficient to keep the carbon in suspension at all times. After two to three days (experimentally determined as the time required for the system to come to equilibrium) the pH was measured and recorded, and an additional quantity of standard acid was added; two to three days later the equilibrium pH was again measured, and another quantity of acid added. This procedure was repeated several times, usually over a period of from two to four weeks, to provide a series of adsorption capacities for decreasing equilibrium pH values. Distilled water was added to the solutions periodically to correct the volume for evaporation. When evaporation was small, it was necessary to correct the data for increased volume owing to addition of acid. The data obtained could then be reduced to moles of acid reacted per gram of carbon at a given hydronium-ion activity, using the extended Debye-Hückel law for calculating activity coefficients. The activity coefficient for the $1M$ NaCl solution, $\gamma = 0.75$, was obtained from Harned and Owen (12). Freundlich parameters were obtained from log-log plots of the experimental data. Except when otherwise specified, the data reported in this paper have been obtained from experiments at $25°C$.

Ash Content Analysis. Ash content can be an important factor in determining the adsorptive behavior of an active carbon. Blackburn and Kipling (2) have demonstrated some of the effects of ash content on the adsorption process. To assess the effects of ash on the interaction of strong acid with active carbon, separate quantities of the experimental coconut carbon were washed with $1 + 1$ hydrochloric acid and with $1 + 1$ glacial acetic acid to reduce the ash content of the carbon. Thirty- to forty-gram samples of carbon were shaken with the acids for about five days. The carbons were then washed continuously with distilled water for a period of three months until the carbon could be contacted with distilled water for a few days without significantly reducing the pH of the water. The carbons were then dried at $105°C.$ to a constant weight. The ash content was measured by burning a known weight of the carbon at $700°C.$ and weighing the residue. The ash content of the carbon washed with acetic acid was reduced from 0.7% to 0.6%, while that washed with hydrochloric acid was reduced from 0.7% to 0.3%. Adsorption studies were then performed on the treated and untreated carbon for purposes of comparison.

Stoichiometry Studies. The HCl–NaCl system was studied to determine if Cl⁻ ion was removed stoichiometrically with H_3O^+ as the acid sorption reaction occurred. Two 1-liter solutions were prepared at pC_{H^+} = 3.00 and $10^{-3}M$ NaCl; two 1-liter solutions at pC_{H^+} = 3.00 and $2 \times 10^{-3}M$ NaCl; and two 1-liter solutions at pC_{H^+} = 2.70 and $10^{-3}M$ NaCl. A five-gram quantity of coconut-shell carbon was added to each of the pC_{H^+} = 3.00 solutions, and a ten-gram quantity to each of the pC_{H^+} = 2.70 solutions. After one day of stirring, the residual H_3O^+ concentration was measured with a pH meter and corrected for activity, and the residual Cl⁻ ion concentration was determined by means of the Mercuric Nitrate Method (24). The percent stoichiometry was then calculated from the data obtained.

Results and Discussion

The rate-limiting step for the hydronium ion-active carbon reaction appears to be intraparticle transport. From a phenomenological point of view, both the hydronium ion and the conjugate anion of the acid added are removed stoichiometrically. The studies made on the HCl–NaCl systems described above show that the ratio of Cl⁻ ion removed to H_3O^+ ion removed is in the range of 0.93:1 to 1:1. This corresponds to similar findings by Carr et al. (4) and Miller (16). The data indicates that the electroneutrality requirement for sorption is satisfied in this reaction. Since the hydronium ion and anion are removed from solution stoichiometrically, they would also be expected to diffuse through the pore in pairs with the anion limiting the rate of diffusion. The diffusion coefficients for different acids, calculated on the basis that intraparticle transport is rate-limiting, are on the order of those expected for the anion, thus giving support to this assumption. Helfferich (13) states that film diffusion, the other likely possibility for being the rate-limiting step, will control only under extreme conditions. A study of sorption rate vs. stirring speed showed no increase in rate for stirring speeds above that required to keep the carbon in suspension. Other evidence in support of intraparticle transport as the rate controlling mechanism is the fact that the experimental data are described well by a diffusion model, as will be illustrated shortly. An activation energy of -1.8 to -2.5 kcal./mole-°K., also discussed in more detail in a later section of this paper, falls in the range expected for a diffusion-controlled process (27).

Diffusion Model. Assuming that pore diffusion is rate limiting, a diffusion model based on Fick's second law can be utilized for calculation of diffusion coefficients from the experimental data. The model must take account of the simultaneous diffusion-reaction process. If the sorption reaction is not taken into account, the calculated diffusion coefficients will deviate considerably from the actual values. Such a model has been developed by Crank (7)—and utilized later by Weber and Rumer (29)—by modifying Fick's second law to include a term which accounts for non-linear sorption. The general form is

$$\frac{\delta C}{\delta t} = \frac{1}{r^2} \frac{\delta}{\delta r}\left(r^2 D \frac{\delta C}{\delta r}\right) - \frac{\delta S}{\delta t} \qquad (1)$$

Equation 1 is given in spherical coordinates, thus assuming a spherical shape for the carbon particle, an assumption which accords reasonably well with microscopic observations of the geometry of particles of the experimental carbon. In Equation 1, C represents the H_3O^+ activity in solution; t, time; r, the radial distance from the particle center; D, the diffusion coefficient; and S, the H_3O^+ concentration at the surface of the carbon. For the present experiments, the equilibrium relationship between S and C is described in terms of the Freundlich expression

$$S = RC^N, N < 1.0 \qquad (2)$$

Equation 1 is subject to the boundary conditions

$$C = 0 \quad t = 0 \quad \text{for} \quad 0 \leqslant r \leqslant a$$

and

$$\frac{A}{\gamma}(C_o - C_a) = 4\pi \int_0^a \left(\frac{C}{\gamma} + S\right) r^2 dr \qquad (3)$$

where A represents the volume of solution "served" by each particle; C_o, the initial H_3O^+ activity in bulk solution; C_a, the H_3O^+ activity in bulk solution at any time; γ, the activity coefficient; and a, the radius of the spherical particle. Equation 1 is a non-linear, partial differential equation which can be solved by a finite difference technique after Crank (7). The solution is made easier by expressing the equation in terms of the dimensionless parameters:

$$c = \frac{C}{C_o}, \pi = \frac{Dt}{a^2}, \rho = \frac{r}{a}, s = \frac{S}{C_o} \qquad (3)$$

Before S can be divided by C_o to give the dimensionless quantity, s, it must be converted from units of moles per gram to moles per liter by multiplying by the specific weight of the carbon. The specific weight for the 273-μ coal-base carbon is 0.750 grams per cc. as given by the manufacturer (20), and that for the coconut-shell carbon is assumed to be the same. The equation requiring solution, in dimensionless form, is then

$$\frac{\delta c}{\delta \tau} = \frac{1}{\rho^2} \frac{\delta}{\delta \rho} \left(\rho^2 \frac{\delta c}{\delta \rho} \right) - \frac{\delta s}{\delta \tau} \quad (4)$$

with the boundary conditions,

$$c = 0 \quad \tau = 0 \quad \text{for} \quad 0 \leqslant \rho \leqslant 1$$

$$\frac{A}{\pi a^2 \gamma} (1 - c_1) = 4 \int_0^1 \left(\frac{c}{\gamma} + s \right) \rho^2 d\rho \quad (5)$$

and,

$$s = R'c^N \quad (6)$$

where R', the dimensionless Freundlich parameter, is given by

$$R' = R \times \text{sp. wt. of carbon} \times C_o{}^{N-1}$$

Assuming a spherical particle and using the specific weight, the term A can be calculated for the finite-bath systems. For the 1.5 gram/liter dosage used in these studies, $A = 0.00533$ cc. for both the coconut-shell and the coal-base carbons.

An IBM 7090 digital computer was used to calculate all rate curves. Calculated curves for the finite bath studies were printed out in the form C_a/C_o vs. τ. These curves were then compared with the experimental C_a/C_o vs. t curves, and D was calculated by a trial and error procedure. This approach was modified slightly in order to calculate the rate curves for the infinite bath studies. The value for "A" taken for calculation purposes, in this case 1000 cc., was very much greater than the actual value, thus producing the effect of an essentially constant C_a with time. In this case, the integral value

$$4\pi \int_0^a \left(\frac{C}{\gamma} + S \right) r^2 dr \quad (7)$$

which represents the total amount of solute in one carbon particle, was printed out as a function of the dimensionless time parameter, τ. This value, when multiplied by the total number of particles present in the solution volume, represents the number of moles of acid added to the solution to maintain a constant pH. By comparing the calculated data to the experimental curves for H_3O^+ added vs. time, the diffusion coefficient could be calculated by a trial and error procedure. Carbon dosages for the finite bath experiments were 1.0 to 2.0 grams per liter.

The basic assumptions made in developing the model should be kept in mind when analyzing the results. Sorption is assumed to be occurring at a rate much faster than pore diffusion, while pore diffusion is taken to be slower than film diffusion. The model assumes radial flux only, and a homogeneous particle. Thus, no differentiation is made between pore volume and solid carbon. Also, the assumption of spherical geometry neglects the fact that the particle surface is extremely irregular. The diffusion coefficient, D, is also assumed constant with varying acid concentration.

The inclusion of the non-linear isotherm effect requires that the isotherm parameters be known in order to calculate the rate curves. Thus, for each system analyzed, both rate and equilibrium data were required.

Hydrogen–Ion Activity. The isotherm shown in Figure 1 was determined, by the procedure described, for the coconut carbon in a $10^{-2}M$ NaCl system at 25°C. The data shown is a composite of several runs. The Freundlich equation, determined from a log-log plot of this data, is $S = 6.5 \times 10^{-4} C^{(0.127)}$.

Finite and infinite bath rate data were collected for the same system for different initial hydronium ion activities. Figure 2 presents the calculated and experimental rate curves for the finite bath experiments and Figure 3 those obtained by the infinite bath method. Table I is a summary of the results.

Table I. Diffusion Coefficients as a Function of Initial pH for the HCl–NaCl System

$10^{-2}M$ $NaCl$, $25°C.$, $S = 6.5 \times 10^{-4} C^{(0.127)}$

Initial pH (HCl)	Diffusion Coefficient, D, $(cm.^2/sec.) \times 10^7$ (Finite Bath)	Diffusion Coefficient, D, $(cm.^2/sec.) \times 10^7$ (Infinite Bath)
3.50	6.75	5.9
3.70	9.50	7.8
4.00	13.00	8.6

The magnitude of the diffusion coefficients given in Table I can be compared with a value of 3.3×10^{-5} cm.^{2}/sec. determined experimentally by Stokes (26) for HCl in bulk solution at infinite dilution. The pore diffusion coefficients listed in Table I for HCl vary by a factor of $(2 - 4) \times 10^{-2}$ from that given by Stokes. McNeill and Weiss (15) have indicated that active carbon can be considered as a weak-base anion-exchange sorbent. According to Helfferich (13), diffusion coefficients in such resins can be several orders of magnitude less than the corresponding bulk solution coefficients. The Cl$^-$ ion probably limits the rate of diffusion, since its mobility in aqueous solution is much less than that of the H_3O^+ ion. Further evidence to support this conclusion has been obtained in the present work from determinations of pore diffusion coefficients for several different strong acids; these will be discussed shortly (see Table IV).

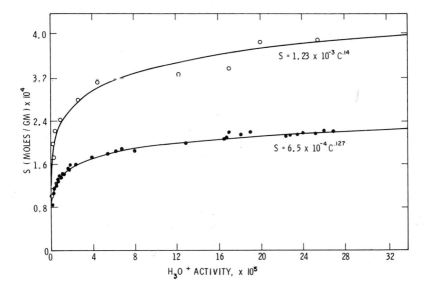

Figure 1. Adsorption isotherms for hydrochloric acid

Adsorption capacities for HCl on coconut-shell carbon (open circles) and on coal-base carbon (solid circles) in 0.01 molar solutions of NaCl are plotted as a function of the hydronium-ion activity at equilibrium. The solid lines represent the calculated Fruendlich isotherms, the equations for which are given on the plot

The large difference between the bulk solution coefficient determined by Stokes and the pore coefficients calculated from the present experiments may be attributable to one or more of a number of factors, the most likely of which is the size and geometry of the pores of the active carbon. The smaller the pore, the more likely is the diffusing ion to interact with internal surfaces, both physically and electrostatically (13). The relative effect of pore size is indicated by comparative studies carried out with 273-micron coal-base carbon. Although there was no pore size distribution data available for the coconut carbon, the average size of the micropores of this carbon can be assumed to be considerably smaller than that of the coal-base carbon, since the former was designed for adsorption from gas phase while the latter was designed for adsorption from solution. The solution used for this equilibrium study had an initial pH of 3.50 (HCl), was $10^{-2}M$ in NaCl, and contained 1.33 grams of the coal-base carbon per liter; the study was made at 25°C. The isotherm for this system is shown in Figure 1, along with the isotherm for the coconut carbon. Kinetic measurements were made on the same system

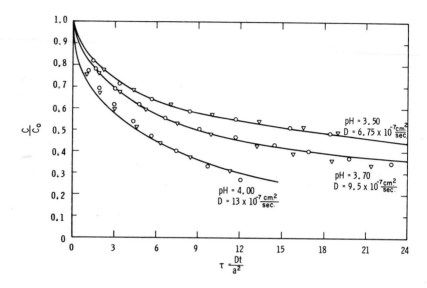

Figure 2. Finite-bath rates of adsorption for hydrochloric acid

The solid lines are calculated rate curves drawn through experimental data for systems of different initial pH. The ordinate is the ratio of acid concentration, and the abscissa is the dimensionless time parameter. Data for two experimental runs are shown for each initial pH; the corresponding diffusion coefficients are noted for each rate curve

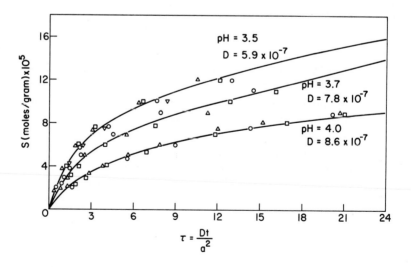

Figure 3. Infinite-bath rates of adsorption for hydrochloric acid

The solid lines are calculated rate curves drawn through experimental data for systems in which pH was held constant over the entire course of each run. Dosages of coconut carbon ranged from 1 to 2 grams per liter. The different symbols represent replicate runs for each pH

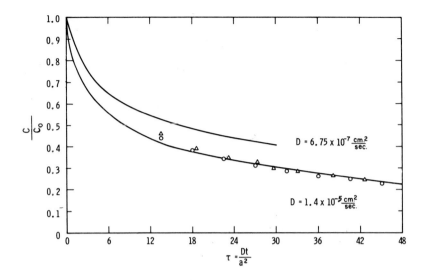

Figure 4. Finite-bath rates of adsorption for hydrochloric acid on different carbons

Experimental data and the corresponding calculated rate curve are shown for the coal-base carbon. The calculated curve for the coconut-shell carbon is reproduced from Figure 2 for purposes of comparison

using the finite-bath technique, and the calculated and experimental rate curves are shown in Figure 4. The corresponding rate curve for the coconut carbon with this system is included in Figure 4. The calculated diffusion coefficient for HCl in the coal-base carbon is 1.4×10^{-5} cm.^{2}/sec., which is nearly equal to that for HCl in bulk solution. Although many other factors might contribute to the large difference between the two carbons, relative pore size is probably a major one.

Other factors which may also be responsible to some extent for the small coefficient obtained for the coconut carbon include the approximation of spherical particle geometry, the assumption of an isotropic medium, and the assumption of a radial diffusion path (*13*).

Specific chemical interaction between the chloride ion and metal ions present at the pore surfaces has also been considered as a possible factor contributing to retardation of HCl diffusion. To evaluate this possibility, one portion of the coconut carbon was washed with acetic acid to reduce its ash content from 0.7% to 0.6%, and another portion with hydrochloric acid to reduce its ash content from 0.7% to 0.3%. The finite bath technique was used to study these two carbons in otherwise identical systems consisting of $10^{-2}M$ NaCl, 1.5 grams carbon per liter, an initial pH of 3.50 (HCl), at a temperature of 25°C. The corresponding isotherms show no significant difference between these carbons and the

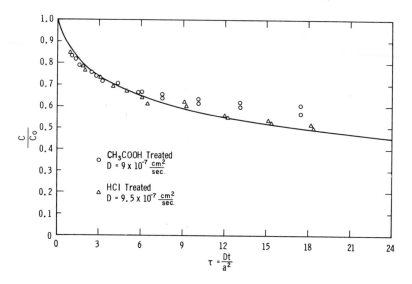

Figure 5. The effect of ash removal on finite-bath adsorption rate

Data are presented for adsorption of HCl on acid-treated coconut carbon. Data for two separate runs are shown for each of the acid-treated carbons. The calculated curve has been derived from the data for adsorption on untreated coconut carbon in the same type of system

untreated coconut carbon. The experimental rate data and calculated curves for the two systems are plotted in Figure 5. Because there was found to be no essential difference between the calculated curves for the acid treated and the untreated carbon, only one curve is shown. The experimental rate data for the acetic-acid-treated carbon are not described as well as are those for the hydrochloric-acid-treated carbon, nor as well as those for the untreated carbon fitted by the calculated curve shown in Figure 1. The diffusion coefficient derived from the curve of best fit for the acetic-acid-washed carbon is 9×10^{-7} cm.^{2}/sec., and that for the hydrochloric-acid-washed carbon is 9.5×10^{-7} cm.^{2}/sec.; this compares with 6.75×10^{-7} cm.^{2}/sec. for the untreated carbon. The increase in the diffusion coefficient for the treated carbon over that for the untreated carbon indicates that interaction of the HCl with the inorganic content of the carbon could be responsible to some extent for retarding diffusion. Analysis has shown that the ash is approximately 50% iron, which in ionic form tends to form complexes with the Cl$^-$ ion.

Diffusion coefficients for different initial pH calculated from data obtained by the finite bath technique compare reasonably well with those determined by the infinite bath technique, as can be seen from the data in Table I. The calculated rate curves fit the experimental data very well for initial pH values of 3.50 and 3.70, but for an initial pH of 4.00, finite-bath case, the fit is not as good. For the latter case, the range of the diffusion coefficient is $(6 - 14) \times 10^{-7}$ cm.^{2}/sec. for all of the experimental data to fall on the calculated curve. Another effect shown by the data is that the diffusion coefficient increases with increasing initial pH, indicating that the coefficient is a function of activity. This is contrary to the assumption of a constant diffusion coefficient built into the rate model. However, except for the case of an initial pH of 4.0, finite bath case, the effect is apparently not very great for the concentration change occurring in any of the rate studies depicted in Figure 2; otherwise the experimental data would not be described as well by the calculated curves. The decrease in diffusion coefficient with increasing H_3O^+ activity is probably caused by hydration effects. The increased number of hydrated H_3O^+ and Cl^- ions in the pore decreases the amount of free solvent available for diffusion of the ions, thus decreasing effective pore size and increasing retardation effects caused by interaction of the diffusing species with the carbon framework. A decrease in the diffusion coefficient is thereby brought about (*13*).

Salt Effects. The HCl–NaCl system has been studied further to determine the effect of NaCl concentration on the acid-carbon reaction. The carbon dosage was varied from 0.33 to 2.0 grams of 273-micron

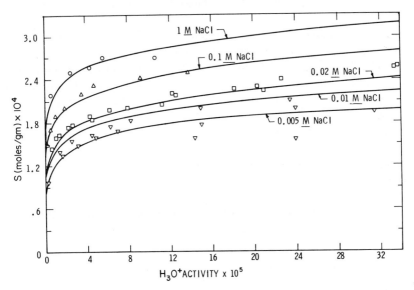

Figure 6. The effect of salt concentration on adsorption capacity for hydrochloric acid

Isotherms are presented for adsorption of hydrochloric acid on coconut carbon in the presence of different concentrations of NaCl. The solid lines represent the corresponding calculated Freundlich isotherms, respective values for the parameters of which are given in Table II

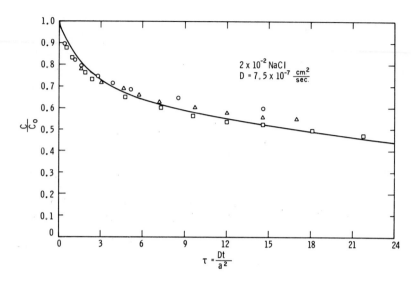

Figure 7. Finite-bath rate of adsorption for hydrochloric acid at higher salt concentration

The concentration of NaCl in the experiment represented by these data was two times that in the experiment for which rate data are given in Figure 2 at pH = 3.5. All other factors were the same

coconut carbon per liter for the equilibrium studies, and the temperature was 25°C. NaCl concentration was studied over a range from 5×10^{-3} to $1.0M$. The effect of salt concentration on the equilibrium position of the reaction is shown in Figures 7 and 9. Typical rate data, along with the corresponding calculated curve are shown in Figure 7 for $2 \times 10^{-2}M$ concentration of NaCl. The finite bath method was used for these rate studies. A summary of the results is given in Table II. It should be noted that the range of the Freundlich parameters used for the rate curve calculations produced very slight changes in the shapes of the resulting curves.

Table II. The Effect of NaCl Concentration on Diffusion of HCl

Initial pH = 3.50, Temperature = 25°C.

Freundlich Parameters		NaCl Concentration	D,
$R, \times 10^4$	N	$(moles/liter) \times 10^2$	$(cm.^2/sec.) \times 10^7$
5.6	0.125	0.5	4.0
6.5	0.127	1.0	6.8
6.9	0.133	2.0	7.5
7.1	0.120	10.0	15.1
7.6	0.105	100.0	68.0

Similar studies have been carried out on the $HClO_4$–$NaClO_4$ system. All conditions were the same as those for the HCl–NaCl tests, except that the carbon dosage was 1.33 grams/liter for the equilibrium study and the concentration of $NaClO_4$ ranged from 5×10^{-3} to $10^{-1} M$. The effect on the extent of reaction is shown in Figures 8 and 9, and typical rate data along with the corresponding calculated curve for a concentration of $10^{-2} M$ $NaClO_4$ are given in Figure 10. The results are summarized in Table III.

Table III. The Effect of $NaClO_4$ Concentration on Diffusion of $HClO_4$

Initial $pH = 3.50$, Temperature $= 25°C$.

Freundlich Parameters		$NaClO_4$ Concentration	D,
$R, \times 10^4$	N	(moles/liter) $\times 10^2$	(cm.^{2}/sec.) $\times 10^6$
7.6	0.1050	0.5	4.7
8.2	0.1030	1.0	5.3
8.4	0.1000	2.0	6.0
10.0	0.1065	10.0	10.0

The equilibrium studies were performed with only sodium salts. Rate studies, however, were performed with LiCl and KCl as well as with NaCl, and no significant difference was found.

There is a marked increase in the quantity of acid sorbed with increasing salt concentration for both the HCl–NaCl and the $HClO_4$–$NaClO_4$ systems. A two-hundred-fold increase in NaCl concentration produces a 60% increase in HCl sorption at pH 3.40, while a twenty-fold increase in $NaClO_4$ concentration causes a 30% increase in $HClO_4$ sorption at the same pH. This observation is consistent with a physical sorption model. Electrophoretic mobility measurements have shown that the active carbon has a negative surface potential. It is possible, therefore, that the proton is primarily adsorbed while the anion is secondarily adsorbed in the double layer. If the proton is adsorbed readily but held back because the anion is not easily taken into the double layer, then an increase in salt concentration would have the effect of increasing the anion pressure, and more acid would tend to be adsorbed. The effect is consistent with that noted by Steenberg (9), and Parks and Bartlett (19). Indeed, Boehm (3) found evidence that physical sorption of acid took place on the basal planes of the microcrystallite. Other sorption mechanisms can not be ruled out, however, because reversible chemisorption could show a similar effect within a given range of salt concentrations. Also, pores which might be inaccessible at low salt concentrations could become of importance at the higher anion pressures.

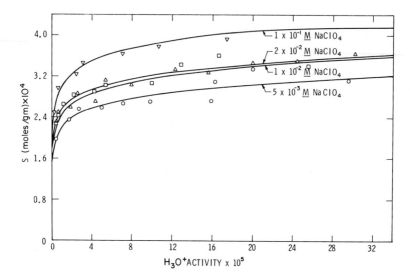

Figure 8. The effect of salt concentration on adsorption capacity for perchloric acid

Isotherms are presented for adsorption of perchloric acid on coconut carbon in the presence of different concentrations of $NaClO_4$. The solid lines represent the corresponding calculated Freundlich isotherms, respective values for the parameters of which are given in Table III

The increase in the diffusion coefficient for the HCl–NaCl system with increased chloride concentration is especially noteworthy. This effect can be best explained by the fact that the driving force for the rate-limiting Cl^- ion is increased, with a resulting increase in the HCl flux. Since the effect of anion concentration is not built into the mathematical diffusion model, higher diffusion coefficients result. The reason for the effect being greater for the HCl–NaCl system than for the $HClO_4$–$NaClO_4$ system may be related to relative ionic size and to the fact that the diffusion coefficient for the HCl is much further from its bulk solution value than is that for $HClO_4$. Because of hydration effects, the Cl^- ion is larger in solution than is the ClO_4^- ion (5). As a result, there are probably many more retardation effects owing to ionic size which can be overcome by the increase in Cl^- ion activity. The fact that the diffusion coefficient for $HClO_4$ is ten times as large as that for HCl could also be attributed to the relative sizes.

The much higher capacity of the active carbon for $HClO_4$ than for HCl has been observed by others for ion exchange resins (5). The observation is in keeping with the smaller hydrated radius of the ClO_4^- ion as noted above. Chu *et al.* (5) have claimed that an additional effect derives from the inability of the ClO_4^- ion to orient surrounding water molecules in bulk solution as effectively as does the Cl^- ion. The water

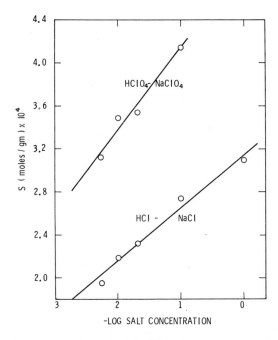

Figure 9. Variation of adsorption capacity with salt concentration

Adsorption capacities for different salt concentrations are shown for the $HClO_4$–$NaClO_4$ and HCl–$NaCl$ systems for an equilibrium H_3O^+ activity of 24×10^{-5}

molecules adjacent to the ClO_4^- ion are attracted more strongly to other water molecules than to the ClO_4^-; thus, a disrupture zone is created in the structure of the bulk-solution water. Because the structure of the water in the resin (active carbon) is already highly disrupted by the charged internal surfaces, the bulk-phase water tends to "push" the ClO_4^- ion into the resin. The Cl⁻ ion, however, tends to be held back in bulk phase because its attraction for the water is stronger than is that of the ClO_4^- ion.

Adsorption of a Series of Strong Acids. Additional kinetic and equilibrium studies have been performed with HNO_3 and H_2SO_4. The results of the equilibrium studies are given in Figure 11; comparable isotherms for HCl and $HClO_4$ are shown also for comparison. These studies were carried out at 25°C. in $10^{-2}N$ solutions of the sodium salt of the conjugate base. Carbon dosages were again 1.33 grams of 273-micron coconut carbon per liter for the equilibrium studies. Typical rate data for HNO_3 in a $10^{-2}N$ $NaNO_3$ solution are shown in Figure 12. Table IV is a tabulation of the results.

Table IV. Diffusion Coefficients for Different Strong Acids

Initial pH = 3.50
Sodium Salt Concentration = 10^{-2}N
Temperature = 25°C.

Acid	Freundlich Parameters		D
	$R, \times 10^4$	N	$(cm.^2/sec.) \times 10^7$
H_2SO_4	14.5	0.1970	2.00
HCl	6.5	0.1270	6.75
HNO_3	6.6	0.1065	24.50
$HClO_4$	8.2	0.1030	53.00

The equilibrium curve for the HNO_3 falls between those for the HCl and the $HClO_4$, as expected on the basis of the explanation given previously for the difference between HCl and $HClO_4$ capacities. The NO_3^- ion would be less hydrated than the Cl^- ion but more hydrated than the ClO_4^- ion. The disrupture effect on the bulk water structure would thus be expected to be intermediate, and the capacity sequence to be $HClO_4$ > HNO_3 > HCl, as observed. This corresponds to the affinity scale given by Kitchner for ion exchange resins (14). Because of the valence effect on the interaction of the SO_4^{2-} ion with the active carbon and on the relative extent of hydration, prediction of H_2SO_4 capacity relative to the monoprotonic acids is difficult. It is interesting to note that the active carbon has a higher capacity for $HClO_4$ than for H_2SO_4. Thus, it is

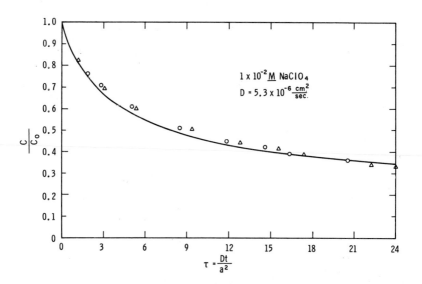

Figure 10. Finite-bath rate of adsorption for perchloric acid

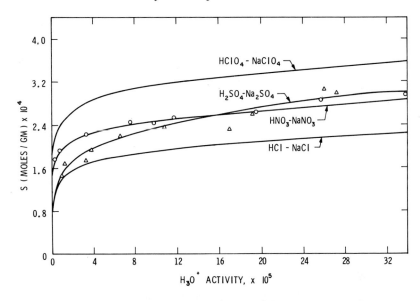

Figure 11. Adsorption isotherms for different strong acids

Data and calculated Freundlich isotherms are presented for HNO_3 and H_2SO_4 for adsorption on coconut carbon from 0.01-normal solutions of $NaNO_3$ and Na_2SO_4, respectively. The isotherms for HCl and $HClO_4$ are reproduced from Figures 1 and 8, respectively, for purposes of comparison

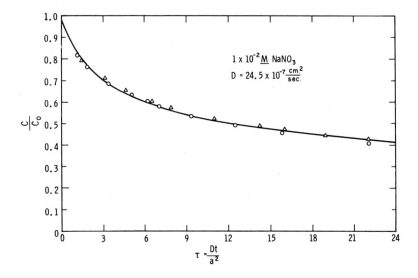

Figure 12. Finite-bath rate of adsorption for nitric acid

probably true that the more extensive hydration of the SO_4^{2-} prevents it from entering some of the smaller pores.

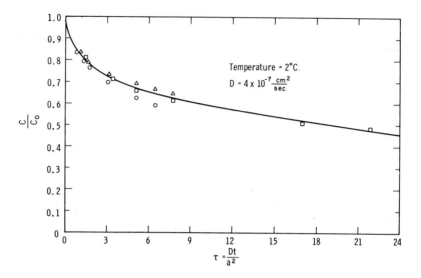

Figure 13. Finite-bath rate of adsorption for hydrochloric acid at 2°C.

The diffusion coefficient for HNO_3 is intermediate between those for HCl and $HClO_4$, in keeping with the relative effects of hydration and disruption of water structure. The diffusion coefficient for H_2SO_4, however, is lower than those of the monoprotonic acids. According to Helfferich (13), the change in the magnitude of the diffusion coefficient between bulk solution and resin phase is much greater for divalent ions than for monovalent ions. This effect could be the result of a combination of increased electrostatic interaction with internal surfaces and of increased physical resistance attributable to the size of the hydrated ion. The sequence of the diffusion coefficients,

$$SO_4^{2-} \ (2 \times 10^{-7}) < Cl^- \ (6.75 \times 10^{-7}) < NO_3^- \ (24.5 \times 10^{-7}) <$$
$$ClO_4^- \ (53 \times 10^{-7}),$$

compares favorably with that found by Soldano and Boyd (22, 25),

$$PO_4^{3-} \ (0.57 \times 10^{-7}) < WO_4^{2-} \ (1.80 \times 10^{-7}) < Cl^- \ (3.54 \times 10^{-7}) <$$
$$BrO_3^- \ (4.55 \times 10^{-7}).$$

The hydrated radii probably follow the same sequence in each case. In addition, the diffusion coefficients for Cl^- are of nearly the same magnitude.

Temperature Effects. Equilibrium studies have been performed on the HCl, $10^{-2}M$ NaCl system at temperatures of 2°, 25°, and 50°C. No significant differences were noted in the resulting isotherms, so that heats of sorption have not been calculated. The results of rate studies on the same systems at the three different temperatures are shown in Figures 13 and 14 for 2° and 50°C., and in Figure 2 for 25°C. The results are tabulated in Table V.

Table V. Diffusion Coefficients for the HCl–NaCl System at Different Temperatures

Initial $pH = 3.50$, 10^{-2}M NaCl

Temperature °C.	Freundlich Parameters		D $(cm.^2/sec.) \times 10^7$
	R, $\times 10^4$	N	
2	4.4	0.093	4.00
25	6.5	0.127	6.75
50	9.3	0.170	8.50

The diffusion coefficients, as expected, increase with increasing temperature. Variation of the diffusion coefficient as a function of temperature can be expressed in terms of the Arrhenius equation, which, in logarithmic form, is

$$\log \frac{D_2}{D_1} = \frac{E_a}{2.303R} \frac{T_2 - T_1}{T_1 T_2} \qquad (8)$$

In Equation 8, D is the diffusion coefficient in $cm.^2/sec.$; E_a is the activation energy in cal./mole-°K.; R is the ideal gas constant (1.987 cal./mole-°K.); and, T is the temperature on the Kelvin scale. A value of $E_a = -(1.8 \text{ to } 2.5)$ kcal./mole-°K., was calculated from the diffusion coefficients listed in Table V, in line with values expected for electrolyte diffusion (13).

Conclusion

Garten and Weiss (8) have postulated that strong acids react with chromene functional groups on the active carbon surface in the following manner:

Chromene (benzpyran)   Carbonium salt

This reaction can be used to explain some of the observations of the present work. The carbonium ion has a high affinity for the OH⁻ ion, as illustrated by a $pK_b = 11$ (23). The carbonium salt will thus hydrolyze readily:

[chemical reaction scheme: carbonium chloride + H₂O → hydroxyl compound + H⁺Cl⁻]

This reaction would account for the fact that carbon which had been contacted with 1 + 1 HCl for five days could be washed continuously with distilled water over a period of three months to remove essentially all of the sorbed acid. This is illustrated by the fact that essentially the same isotherm was obtained for 1 + 1 HCl-treated carbon as for untreated carbon.

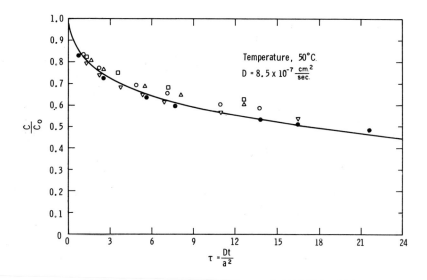

Figure 14. *Infinite-bath rate of adsorption for hydrochloric acid at 50°C.*

The possibility that diffusion of oxygen into the pore is rate limiting has been ruled out because the solutions used were exposed to the atmosphere, and oxygen has been found to be rate limiting only when the partial pressure is less than 20 mm. Hg (8).

It has been noticed also that only a small fraction of the total amount of sorbed acid could be removed by washing with water which contained the salt of the acid in the same concentration as was present in the solution with which the carbon was equilibrated originally. This observation suggests that the conjugate anion is tightly held, and is thus consistent with the findings of others (9, 18, 30).

It would appear, however, that not all the acid is taken up by means of the chromene reaction, since 50 to 60% of the acid sorbed at an equilibrium pH of approximately 3.5 can be displaced by adding phenol to the solution in sufficient quantity to bring the phenol concentration to approximately $0.1M$ at equilibrium. This percentage, at least, of the acid originally sorbed is evidently taken up as a displaceable entity. Garten and Weiss (9) have shown that the amount of acid which can be displaced by a strongly adsorbing organic molecule is dependent upon the extent of oxidation of the active carbon during manufacture. The nature of reactions at the surface therefore depends upon the method of preparation. It becomes difficult to make any generalized conclusions regarding the displaceable character of the sorbed acids. It should be noted that some of the sorbed acid may be in pores which are not readily accessible to the organic molecule and thus are not displaced by the latter.

There is not sufficient evidence conclusively to prove or disprove one or two definite mechanisms for reaction of acid with active carbon. However, the chromene-acid reaction as herein described appears to be a logical and significant part of the overall reaction. Physical, electrostatic, or other mechanisms for acid sorption probably account for the remainder of the reaction.

Regardless of the actual mechanism, the rate and extent of the acid reaction with active carbon is an important factor and should prove useful for interpreting sorption data of various kinds. Systematic sorption studies should lead ultimately to a better understanding of the nature of the acid-carbon reaction.

Acknowledgments

The research reported herein has been supported in part by Research Grant WP-00706 from the Federal Water Pollution Control Administration, U.S. Department of The Interior, and in part by a National Science Foundation Traineeship Number GE4802 to Vernon L. Snoeyink.

Literature Cited

(1) Bartell, F. E., Miller, E. J., *J. Am. Chem. Soc.* **44**, 1866 (1962).
(2) Blackburn, A., Kipling, J. J., *J. Chem. Soc.* **1955**, 4103.
(3) Boehm, H. P., "Advances in Catalysis and Related Subjects," Vol. 16, p. 179; Eley, D. D., Pines, H., Weisz, P. B., Eds., Academic Press, New York and London, 1966.
(4) Carr, C. W., Freundlich, H., Sollner, K., *J. Am. Chem. Soc.* **63**, 693 (1941).
(5) Chu, B., Whitney, D. C., Diamond, R. M., *J. Inorg. Nucl. Chem.* **24**, 1405 (1962).

(6) *Columbia Activated Carbon circular*, National Carbon Co., New York, N. Y.
(7) Crank, J., "The Mathematics of Diffusion," Clarendon Press, London, 1965.
(8) Garten, V. A., Weiss, D. E., *Australian J. Chem.* **10**, 309 (1957).
(9) Garten, V. A., Weiss, D. E., *Rev. Pure Appl. Chem.* **7**, 69 (1957).
(10) Garten, V. A., Weiss, D. E., Willis, J. B., *Australian J. Chem.* **10**, 295 (1957).
(11) Gould, J. P., Master's Thesis, Dept. of Civil Engineering, Univ. of Michigan (1967).
(12) Harned, H. S., Owen, B. B., "The Physical Chemistry of Electrolyte Solutions," p. 593, Reinhold Publishing Corp., New York, 1958.
(13) Helfferich, F., "Ion Exchange," p. 234, McGraw-Hill Book Co., Inc., New York, 1962.
(14) Kitchner, J. A., "Modern Aspects of Electrochemistry," No. 2, Bockris, J., ed., Butterworths Scientific Publications, London, 1959.
(15) McNeil, R., Weiss, D. E., "Proc. of the Fourth Carbon Conference," p. 281, Pergamon Press, London, 1960.
(16) Miller, E. J., *J. Am. Chem. Soc.* **46**, 1150 (1924).
(17) Miller, E. J., *J. Am. Chem. Soc.* **47**, 1270 (1925).
(18) Miller, E. J., *J. Phys. Chem.* **36**, 2967 (1932).
(19) Parks, L. R., Bartlett, P. G., *J. Am. Chem. Soc.* **49**, 1698 (1927).
(20) *Pittsburgh Activated Carbon Circular* **#AC7-556**, Pittsburgh Coke and Chemical Co., Pittsburgh, Pa.
(21) Snoeyink, V. L., Weber, W. J., Jr., *Envir. Science Tech.* **1**, 228 (1967).
(22) Soldano, B. A., Boyd, G. E., *J. Am. Chem. Soc.* **75**, 6099 (1953).
(23) Sondheimer, E., *J. Am. Chem. Soc.* **75**, 1507 (1953).
(24) Standard Methods, 11th edition, p. 79, American Public Health Association, Inc., New York, 1960.
(25) Stern, K. H., Amis, E. S., *Chem. Rev's.* **59**, 1 (1959).
(26) Stokes, R. H., *J. Am. Chem. Soc.* **72**, 2243 (1950).
(27) Walker, P. L., Jr., Austin, L. G., Nandi, S. P., "Chemistry and Physics of Carbon," Vol. 2, P. L. Walker, Jr., Ed., Marcel Dekker, Inc., New York, 1966.
(28) Weber, W. J., Jr., Morris, J. C., *J. Sanit. Eng. Div. Amer. Soc. Civil Engrs.* **90**, SA3, 79 (1964).
(29) Weber, W. J., Jr., Rumer, R. R., Jr., *Water Resources Research* **1** (3), 361 (1965).
(30) Weller, S., Young, T. F., *J. Am. Chem. Soc.* **70**, 4155 (1948).

RECEIVED November 24, 1967.

11

Adsorption and Wetting Phenomena Associated with Graphon in Aqueous Surfactant Solutions

F. G. GREENWOOD, G. D. PARFITT, N. H. PICTON, and D. G. WHARTON

University of Nottingham, Nottingham, England

> *Adsorption isotherms at 25° were determined for sodium dodecyl sulfate on Graphon from aqueous solutions and from 0.1M sodium chloride, also for dodecyl trimethylammonium bromide using potassium bromide as electrolyte. The dispersibility of the powder in the various solutions was assessed by measuring the optical density of dispersions resulting from end-over-end shaking. Comparison with coagulation studies confirms that dispersibility is not controlled by the electrochemical properties of the system. Contact angle measurements were made to assess the wetting characteristics of the systems. A fairly discrete value of surface coverage of surfactant ions is necessary before the powder is readily dispersed by end-over-end action. This value is independent of ionic strength, and corresponds to the solution concentration for which the contact angle becomes $< 90°$.*

The problem of incorporating a powder into a liquid to form a dispersion of fine particles is an important aspect of colloid chemistry. The overall process may be considered as consisting of three stages:

1. Wetting of the powder. Powders consist of aggregates and agglomerates (two ways of defining clusters of primary particles (9)) so not only the wetting of the external surfaces but also the displacement of air and wetting of the internal surfaces (between the particles in the clusters) must be considered. The effectiveness of the wetting process may be expressed in terms of the solid/liquid/vapor contact angle which must be zero for spontaneous wetting of the external surface (*14*), and less than 90° for spontaneous penetration into the agglomerates (*1*).

2. Breaking up the aggregates and agglomerates into colloidal particles. Ideally the work required to complete this stage should be as small as possible, although in some cases large energies may be involved depending on the strength of the bond holding the primary particles together in the clusters. For the systems considered in this paper little effort is apparently required for this stage. It has been suggested (16) that the resistance to stress of particle-particle bonds can be significantly reduced by the addition of surface active material but the mechanism of the process is not established.

3. Coagulation (reduction in particle number with time due to irreversible collisions) of the dispersion. The resistance to coagulation, or the stability of the dispersion, depends on the relative magnitudes of the attractive van der Waals forces between the particles, and the repulsive force which in a system involving charged particles may be associated with the overlapping of their electrical double layers. The stability of a colloidal dispersion is predicted by the Deryaguin-Landau-Verwey-Overbeek (DLVO) theory (6, 7, 21).

Dispersibility has been defined (12) as the ease with which a dry powder may be dispersed in a liquid and this term can be used to express the effectiveness of the first two stages. Although in theory the three stages may be considered quite separately, interpretation of experimental observations in terms of these stages may be difficult because they usually overlap in practice. A great deal of attention has been paid to the factors involved in the stability of colloidal dispersions in relation to current theories. The relationship between dispersibility and the various parameters obtaining in any particular system has received little attention. The wetting characteristics of aqueous surfactant solutions on oxide etc. surfaces is of considerable interest to mineral processing, and on carbon blacks to detergency, but surprisingly few attempts have been made to relate the efficiency of the processes to the interfacial tensions prevailing and to the contact angles. Unfortunately the measurement of contact angle for a liquid with a powder is beset with difficulties.

The adsorption of the surface active agent at the solid/liquid interface is, presumably, an important prerequisite to the process associated with dispersibility. Besides the lowering of the interfacial tension, another factor is involved with ionic agents namely the electric potential associated with adsorption of ions. Both factors were considered by Tamamushi (19) to be relevant to the dispersion of powders in aqueous surfactant solutions. Some authors (15, 20, 22, 23) have related the effects directly to the zeta potential, while others (8, 11, 18) discuss their observations in terms of the increasing degree of hydrophilic character of the carbon black surface as a result of adsorption. Fundamental energetic considerations show (12) that the values of the contact angle and the surface tension of the wetting liquid are important parameters controlling the dispersion process. The effects of coagulation, controlled by the electrical properties of the system, may be superimposed on the dispersing process and this may lead to an incorrect interpretation of the experimental results.

This paper describes a study of the dispersibility of Graphon (graphitized Spheron 6) in aqueous solutions of sodium dodecyl sulfate(SDS) an dodecyl trimethylammonium bromide (DTAB), and its relation to the adsorption behavior of the surfactants at the solid/liquid interface, with a view to determine the controlling process in the dispersibility of these systems.

Experimental

Materials. Graphon (the graphitized form of the medium-processing channel black, Spheron 6) was supplied by the Cabot Corporation. The surface area of Graphon (4) of 78.9 meter^2/gram was determined by the B.E.T. method using nitrogen at $-196°C$. and $\sigma = 16.2$ A.^{2}. Pure samples of DTAB and SDS were supplied by Glovers Chemicals Ltd. and Cyclo Chemicals respectively. Analysis of the surfactants gave the following results: SDS, C 49.76% (calc. 50.00%), H 8.73% (calc. 8.68%), residue 25.12% (calc. 24.66%) and $> 99\%$ C_{12} homologue; DTAB, N 4.40% (calc. 4.54%), Br 25.44% (calc. 25.92%) residue 70.16% (calc. 69.54%) and $> 96\%$ C_{12} homologue. Values of the critical micelle concentration (c.m.c.) were determined for the two surfactants and the results for SDS, c.m.c. = 8.0 mM. (drop volume method for surface tension; no minimum observed), and DTAB, c.m.c. = 16.0 mM. (conductance), were in good agreement with literature values (17, 24), indicating a satisfactory level of purity. B.D.H. Ltd. cetyl pyridinium bromide (standard cationic agent), and A.R. sodium chloride and potassium bromide were used.

Procedure. For the adsorption measurements samples of about 0.3 gram Graphon were accurately weighed into adsorption tubes, about 10 ml. of surfactant solution added, the tubes sealed and rotated end-over-end in a water thermostat at 25 ± 0.1° for at least twelve hours; it had been established that a much shorter time was required for reaching adsorption equilibrium. Usually it was necessary to separate the solid from the solution by filtering through an Oxoid membrane filter but where possible centrifuging at 3500 r.p.m. was used. The clear supernatant liquid was analyzed for surfactant by titration (2) against cetyl pyridinium bromide for DTAB. Bromophenol blue was used as indicator. In the cases where both separation techniques were available identical results were obtained. The effect on the adsorption of breaking up the Graphon by irradiation with ultrasonics was assessed in similar experiments in which the mixtures were subjected to 40 kc./sec. radiation for two minutes using a 500 watt Dawe Instruments Ltd. Soniclean Generator.

For the assessment of dispersibility, samples of about 0.1 gram of Graphon were accurately weighed into standard tubes approximately 1.5 cm. wide and 13 cm. long fitted with B14 Quickfit joints. A known amount of solution (∼10 ml.) was added and the tubes were rotated end-over-end in the thermostat at 25° at approximately 20 r.p.m. for various times, after which they were allowed to stand for 18 hours so that the larger Graphon particles would settle. The optical density of the remaining dispersion was measured at 400 mμ in a 2 mm. cell using a Unicam SP 600 spectrophotometer, in a constant temperature room maintained at 25 ± 1°. Optical densities were corrected to a concentration of 1 mg. Graphon/ml. solution.

The wetting characteristics of the systems were assessed by measuring the contact angles (θ) of the solutions on the powder using the Bikerman method (3). To obtain a non-porous flat surface on which to place drops of liquid for measurement, a thin layer of Graphon was pressed on a flat paraffin wax surface. Drops of different volumes (0.001 to 0.008 cc.) were placed on the Graphon surface using an Agla syringe, and θ calculated from measurements, using a travelling microscope, of the diameters of the areas of contact of the drops (extrapolated to zero volume), assuming each drop to have the same shape as a segment of a sphere. In cases where this condition was not fulfilled—*i.e.*, at low θ—anomalous results were obtained. Contact angles for water were also measured for Graphon pressed on a vinyl plastic tile and also on a sheet of Polythene, and within experimental error (±2%) the results were the same as those for Graphon on the wax surface. Contact angles of various DTAB solutions on the wax surface were found to be about 30° lower than the corresponding values obtained for Graphon pressed on the wax surface.

Results and Discussion

The adsorption results for SDS on Graphon from aqueous and 0.1M sodium chloride solutions are shown in Figure 1. In both cases saturation adsorption is reached at the c.m.c., the effect of added salt being to decrease the c.m.c. and to increase the maximum adsorption level such that the average area per adsorbed DS$^-$ ion decreases from 42A.^{2} to 33A.^{2}. For aqueous solutions a marked point of inflection is observed at about half the c.m.c., which may indicate a change in orientation, from parallel to perpendicular, of the adsorbed ion. At the point of inflection the area per adsorbed ion is approximately 70A.^{2} which would satisfy the parallel orientation model. Similar experiments (4) on heat-treated samples of the original carbon black Spheron 6 indicate that the point of inflection is associated with the graphitized, homogeneous surface containing virtually no hydrophilic sites. The point of inflection is not apparent in the isotherm for 0.1M sodium chloride possibly because of the steep rise in adsorption at low concentration. The effect of subjecting the Graphon to ultrasonic radiation is to increase slightly the adsorption at concentrations above the point of inflection. Whether this increase may be correlated with a change in the wetting characteristics of the system is uncertain.

Figure 2 shows the adsorption data for DTAB, which have some similarities to those of SDS in that saturation is reached at the c.m.c. and

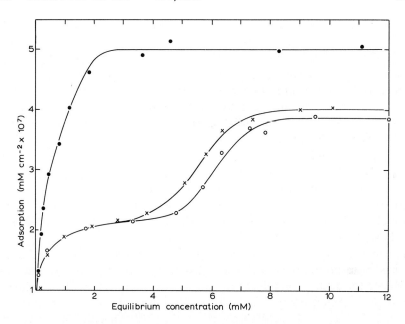

Figure 1. Adsorption of SDS on Graphon at 25° from aqueous solution after end-over-end action ⊙ and after ultrasonic irradiation ×, and from solutions in 0.1M sodium chloride ● (end-over-end)

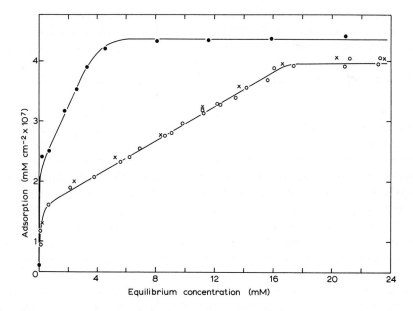

Figure 2. Adsorption of DTAB on Graphon at 25° from aqueous solution after end-over-end action ⊙ and after ultrasonic irradiation ×, and from solutions in 0.1M potassium bromide ● (end-over-end)

the adsorption increases on addition of electrolyte. However, the increase is not as large for DTAB, the average area per DTA$^+$ ion decreasing from 42A.^{2} to 38A.^{2}. This behavior is paralleled by the smaller apparent increase in micellar weight on addition of potassium bromide to DTAB(5) compared with that for sodium chloride on SDS(10), and may well reflect the screening of the nitrogen by methyl groups in DTAB. Subjecting the Graphon to ultrasonic radiation has no effect (within experimental error) on the adsorption.

There are differences in isotherm shape, and for DTAB the behavior is not amenable to a simple explanation. Of particular interest are plots of the amount adsorbed against the mean ionic activity of the surface active agent (including the counterion of the added electrolyte). In the case of DTAB all the data, including others at various salt concentrations up to 0.5M, lie on one line which, after an initial steep rise, is linear to the c.m.c. This indicates that for other than the initial strong adsorption at low concentrations (possibly because of specific interactions with the surface) the adsorption follows the law of mass action. For SDS a similar result is obtained except that positive deviations from the straight line occur below $a_{\pm} \sim 4 \times 10^{-3}M$ for the cases (salt concentration $< 0.1M$) when there is a point of inflection in the isotherm. These deviations may reflect specific interactions of the DS$^-$ with the surface when the ions are adsorbed in parallel orientation.

During the adsorption experiments greater difficulty was experienced in separating the Graphon dispersed in surfactant solutions at concentrations above the c.m.c., and this difficulty increases in magnitude with the length of the period subjected to end-over-end action. Unless the separation is complete, a maximum in the adsorption isotherm is observed since the total amount of surfactant analyzed is larger than that corresponding to the true adsorption. We found the amount of solid remaining suspended that would lead to an adsorption maximum, to be deceivingly small. An illustration of the relation between the equilibrium concentration of DTAB and the amount of solid material remaining suspended after standing for some days following end-over-end action for 30 hours, is given in Figure 3. The initial change, which is fairly abrupt, occurs at a concentration below the c.m.c., and comparison with the adsorption isotherm in Figure 2 shows there to be no apparent correlation between the effect and the nature of the adsorbed layer. Such is the case for all the systems discussed in this paper. Furthermore, the effect bears no relation to the stability to coagulation of the systems. Measurements (13) of the rate of coagulation of dispersions prepared using ultrasonic irradiation show that the Graphon, once dispersed, is indefinitely stable in aqueous surfactant solutions at all concentrations. It is the same for solutions containing salt although in these cases at low

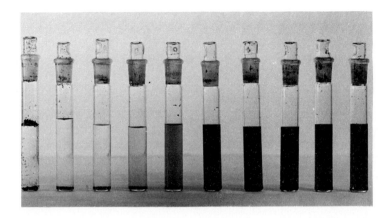

Figure 3. The dispersibility of Graphon in aqueous solutions of DTAB at concentrations from left to right, in mM: 2.0, 5.1, 9.1, 12.3, 15.0, 19.3, 22.6, 26.0, 30.2, 33.7

surfactant concentrations the dispersions are relatively unstable. Also, measurements of electrophoretic mobility indicate that for all the systems the zeta potential is constant over the range of concentration at which there is a marked change in dispersibility. It seems clear that the dispersibility of the Graphon is not controlled by the electrochemical properties of the system.

The dispersibility of Graphon in SDS solutions, both with and without sodium chloride, is illustrated in Figure 4 in terms of the optical density (on a standard weight basis) of the dispersions which remain after various periods of end-over-end action. Similar plots were obtained for DTAB. For all the plots extrapolation to zero optical density of the approximately linear region of rapidly increasing optical density leads to a fairly discrete value of the surface coverage of adsorbed ions (46 ± 1A.^{2} DS$^-$ and 52 ± 1A.^{2} for DTA$^+$) at which the abrupt change in dispersibility occurs. These data indicate that the dispersibility of Graphon is related to the hydrophilic character of the surface associated with the adsorbed surfactant ions.

Spontaneous wetting of the external surface of a solid is associated with zero contact angle, otherwise some work is necessary for complete wetting to be achieved. In the case of a powder we must also consider the penetration of liquid into the small channels inside and between the aggregates of the dry powder, and this is theoretically spontaneous only when $\theta < 90°$ (assuming a hypothetical cylindrical pore). It may therefore be assumed that for the powder to be dispersed in the liquid as fine particles it is necessary for $\theta < 90°$, and that only when $\theta = 0$ would we expect the whole wetting process to be spontaneous—*i.e.*,

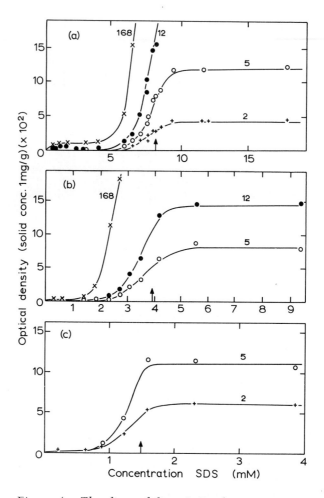

Figure 4. The dispersibility of Graphon in aqueous solutions of SDS shown as values of optical density put on a standard weight basis: (a) without sodium chloride, (b) with 0.02M sodium chloride, (c) with 0.1M sodium chloride. The numbers indicate the number of hours subjected to end-over-end action. Arrows indicate the c.m.c.

require no external work. That this is the case with the present system is borne out by the measurements of θ for the various solutions on Graphon (Figure 5). In all cases it is observed that the Graphon cannot be dispersed easily unless $\theta < 90°$.

It seems reasonable to conclude, therefore, that dispersibility is related to the wetting of the powder by the liquid rather than to the electrochemical properties of the system.

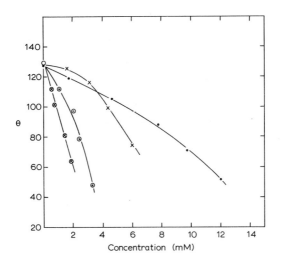

Figure 5. Contact angles on Graphon for aqueous solutions of DTAB ● and SDS ×, for DTAB in 0.1M potassium bromide ⊙, and for SDS in 0.1M sodium chloride ○

Acknowledgments

The authors are grateful to C. D. Moore of Glovers Chemicals Ltd. for preparing the sample of DTAB, to Acheson Industries (Europe) Ltd. for a grant to D.G.W. and to the Research Association of British Paint, Colour and Varnish Manufacturers for a grant to N.H.P.

Literature Cited

(1) Adamson, A. W., "Physical Chemistry of Surfaces," p. 475, 2nd ed., John Wiley and Sons, New York, 1967.
(2) Barr, T., Oliver, J., Stubbings, W. V., *J. Soc. Chem. Ind.* **67**, 45 (1948).
(3) Bikerman, J. J., *Ind. Eng. Chem.* **13**, 443 (1941).
(4) Day, R. E., Greenwood, F. G., Parfitt, G. D., *4th Int. Cong. Surface Active Substances* **Vol. 2**, 1005 (1964).
(5) Debye, P., *Ann. N.Y. Acad. Sci.* **51**, 575 (1949).
(6) Deryaguin, B. V., *Trans. Faraday Soc.* **36**, 203 (1940).
(7) Deryaguin, B. V., Landau, L. D., *Acta Physicochim. U.S.S.R.* **14**, 633 (1941).
(8) Doscher, T. M., *J. Coll. Sci.* **5**, 100 (1950).
(9) Gerstner, W., *J. Oil and Colour Chem. Assoc.* **49**, 954 (1966).
(10) Huisman, H. F., *Proc. Kon. Ned. Akad. van Wet.* **67**, 367 (1964).
(11) Meguro, K., *J. Chem. Soc. Japan (Ind. Chem. Sect.)* **58**, 905 (1955).
(12) Parfitt, G. D., *J. Oil and Colour Chem. Assoc.* **50**, 822 (1967).
(13) Parfitt, G. D., Picton, N. H., *Trans. Faraday Soc.* **64**, 1955 (1968).
(14) Patton, T. C., "Paint Flow and Pigment Dispersion," Chapt. 8, Interscience, New York, 1964.
(15) Ray, L. N., Hutchinson, A. W., *J. Phys. Coll. Chem.* **55**, 1334 (1951).
(16) Rehbinder, P., *Colloid J. U.S.S.R.* **20**, 493 (1958).

(17) Scott, A. B., Tartar, H. V., *J. Am. Chem. Soc.* **65,** 692 (1943).
(18) Tamaki, K., *J. Japan Oil Chem. Soc.* **9,** 426 (1960).
(19) Tamamushi, B., "Colloidal Surfactants," p. 244, Shinoda *et al.,* eds., Academic Press, London, 1963.
(20) Urbain, W. M., Jensen, L. B., *J. Phys. Chem.* **40,** 821 (1936).
(21) Verwey, E. J. W., Overbeek, J. Th. G., "Theory of the Stability of Lyophobic Colloids," Elsevier, Amsterdam, 1948.
(22) Vold, R. D., Greiner, L., *J. Phys. Coll. Chem.* **53,** 67 (1949).
(23) Vold, R. D., Konecny, C. C., *J. Phys. Coll. Chem.* **53,** 1262 (1949).
(24) Williams, R. J., Phillips, J. N., Mysels, K. J., *Trans. Faraday Soc.* **51,** 728 (1955).

RECEIVED October 26, 1967.

12

Analysis of the Composite Isotherm for the Adsorption of a Strong Electrolyte from Its Aqueous Solution onto a Solid

KARMA M. VAN DOLSEN and MARJORIE J. VOLD

The University of Southern California, Los Angeles, Calif. 90007

> *The charge density, Volta potential, etc., are calculated for the diffuse double layer formed by adsorption of a strong 1:1 electrolyte from aqueous solution onto solid particles. The experimental isotherm can be resolved into individual isotherms without the common monolayer assumption. That for the electrolyte permits relating Guggenheim-Adam surface excess, double layer properties, and equilibrium concentrations. The ratio: σ_0/Γ_2^N declines from two at "zero" potential toward unity with rising potential. Unity is closely reached near $kT/e = 10$ for spheres of 1000 A. radius but is still about 1.3 for plates. In dispersions of Sterling FTG in aqueous sodium β-naphthalene sulfonate a maximum potential of $kT/e = 7$ (170 mv.) is reached at 4×10^{-3}M electrolyte. The results are useful in interpretation of the stability of the dispersions.*

It has been recognized since the nineteenth century that sparing soluble particles of large surface area have been known to be maintained for long times in solution of sufficiently low ionic strength. The mutual repulsion of the like charge borne by the particles counterbalances the van der Waals attraction. Two classes of solid/liquid interface have been extensively studied: the completely polarized (3) and the completely reversible interface (9, 12, 13). The origin of the influence of electrolyte is, as is well known, partial expulsion (negative adsorption) of similions near the surface and concomitant increase in concentration of counterions. The situation was treated as analogous to a condenser with the charge per unit area of the colloidal surface and the equivalent net charge in the surrounding solution being designated "electric double layer" as early as 1879 by Helmholz. Shortly thereafter the distribution of both similions and counterions in the potential field was treated by the statistical Boltzmann expression by Guoy and Chapman (5) to allow for the thermal motion of the ions. The concept of capacity of the electric double layer remained, apart from the designation "diffuse double layer."

For a reversible interface, such as AgI/aqueous solution, the electrostatic potential in the solution just outside the surface referred to zero at regions of solution infinitely remote from colloidal particles, the Volta potential, is calculated from the Nernst equation, the concentration of potential determining ions, and the zero-point-of-charge which is not usually the stoichiometric equivalence point.

The characteristics of the diffuse electric double layer at a completely polarized interface, such as at a mercury/aqueous electrolyte solution interface are essentially identical with those found at the reversible interface. With the polarizable interface the potential difference is applied by the experimenter, and, together with the electrolyte, specifically adsorbed as well as located in the diffuse double layer, results in a measurable change in interfacial tension and a measurable capacity.

It can be observed that the above two types of electric double layer, which have basically similar properties, differ principally in the manner of establishing the potential difference across the electric double layer. One type is fixed by the solubility and other interactions of the solid in contact with solution of electrolyte. In the other type, polarizable interface, the experimenter applies any desired potential difference between one liquid surface and a reference electrode. The resulting Volta potential is fixed by the specific adsorbability of the electrolyte.

A third type of electric double layer, hitherto but slightly investigated in modern times, derives its Volta potential from the specific adsorbability of an ion which is chemically unrelated to the solid. This type, which is the subject of the present paper (using beta-naphthalene sulfonate ion and a homogeneous nonpolar graphitized carbon) also has qualitative similarities to the two classical ones. The results of this study are intended for use in companion studies of the electrokinetic properties and stability factors of the same system. Although graphitized carbon black (and also carbonaceous material of varying, sometimes unknown composition) has been investigated as sols in hydrocarbon media (largely as a result of its importance to the petroleum and related industries), and to a lesser extent as aqueous sols to simulate "dirt" in studies of detergency, precise data and careful interpretation of them are both lacking.

One of the most significant results is that negative adsorption of similions (their partial expulsion from the surface region), which has been recognized but not emphasized, even in the earliest days of Guoy-

Chapman theory, is in fact a quantity of considerable importance. Already van den Hul and Lyklema (18) had realized this situation. They showed that the measured negative adsorption of phosphate ions and of sulfate ions could be used to determine the surface area of the particles in negative sols of AgI, in reasonable agreement with results obtained from capacitance measurements. In the present case, a first integration of the Poisson-Boltzmann equation gives the charge density at the surface corresponding in position to that of the Volta potential. But the adsorption measured by change in concentration of sodium beta-naphthalene-sulfonate refers to the number of ions on the surface minus the number of ions which are negatively adsorbed in the diffuse double layer. The two quantities are not identical and differ by a factor of two, if the (dubiously justifiable) Debye-Hückel approximation is used. Happily, the ratio between negative adsorption and total surface charge appears

---

Symbol Table

a	Radius of a sphere (meters).
A	Area of the surface (meters^2).
C^b	Molar concentration of electrolyte in the bulk solution.
e	Charge of an electron (e.s.u.).
k	Boltzmann constant.
m	Weight of adsorbent (grams).
$n_1{}^s$	Moles of component i in the surface-containing region.
N_o	Total moles of all components present in the solution before dispersing the adsorbent.
N	Avogadro's number.
l	Distance (meters).
L	Linear extent of the surface-containing region measured perpendicularly outward from the interface (meters).
r	Distance from the center of a sphere (meters).
T	Absolute temperature.
u	Reduced potential, $z_i e\psi/kT$.
u_o	Reduced surface potential.
$\bar{v}_i$	Partial molecular volume of component i (milliliters/molecule).
V^s	Volume of the surface-containing region (meters^3).
x_i	Mole fraction of component i.
z_i	Valence of the ith ion type.
$\Gamma_2{}^{(N)}$	Guggenheim-Adam "N convention" surface excess of component i (i = 1 for water, i = 2 for electrolyte, in moles/meter^2).
ϵ	Bulk dielectric constant of water.
σ_o	Charge density on the surface (moles/meter^2).
Σ	Specific surface area of adsorbent (meters^2/gram).
ψ	Electrostatic potential (millivolts).
ν_i	Concentration of ions (ions/ml. i = c for counterions, 1 = s for similions, i = b for both in the bulk solution).

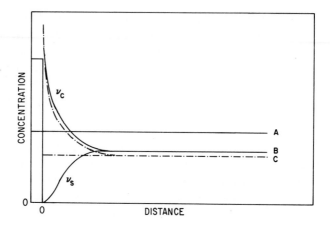

Figure 1. Schematic: Diffuse double layer formed as a result of anion adsorption, showing the effect of negative adsorption of similions on the bulk concentration. A. Initial electrolyte concentration. B. Final electrolyte concentration. C. Final concentration if there were no negative adsorption of similions

(*19*) to be independent of electrolyte concentration and depends only on the Volta potential.

Theory

The presence of an adsorbed layer at the solid/solution interface is inferred from an observed concentration change when colloidal solid is placed in contact with solution. If one of the ions adsorbs preferentially, the interface becomes charged, and a diffuse double layer is formed spontaneously extending outward from the charged surface. Since counterions are attracted to the charged interface, their concentration in the diffuse double layer is greater than in the bulk solution and they are said to be positively adsorbed. Similions are repelled from the surface and are said to be negatively adsorbed. The neutralization of the surface charge is accomplished by the combination of the positive adsorption of counterions and the negative adsorption of similions. Since positive adsorption is inferred from a decrease in concentration of the bulk solution, and negative adsorption is inferred from an increase in concentration, the net change will be smaller in magnitude than the larger of the two effects. The situation is schematized in Figure 1.

This measured concentration change is usually converted into a surface excess quantity, analogous to that usually calculable from surface tension data for adsorption at the liquid/vapor interface. In the case of the solid/liquid interface in which the solid is impenetrable to solvent and solute alike, the surface of contact between the solid and solution is a logical choice for the dividing surface. An arbitrary surface must be chosen to separate the bulk solution from the surface-containing region. All extensive quantities of the interfacial region will be dependent on the position of this second surface. Any convention chosen should satisfy the following criteria: (1) give a value of the surface excess which is independent of the choice of position of the second dividing surface; (2) clearly show the relative presence of both solvent and solute in the interfacial region; and (3) allow a concept of this region which facilitates application of diffuse double layer theory.

The Gibbs surface excess, $\Gamma_2^{(1)}$, is conceptually difficult, and has the further disadvantage that the extent of the surface region must change as the composition of the surface region changes (4). The Guggenheim-Adam "N" convention surface excess, Γ_2^N, is a logical choice inasmuch as it satisfies the above criteria and also allows resolution of the composite isotherm into individual isotherms (8). The Guggenheim-Adam surface excess is defined as the number of moles of electrolyte in a volume of solution containing one square meter of surface and N total moles of all species, in excess of the corresponding quantity in a volume of bulk solution containing the same total number of moles (4). This surface excess quantity is symmetrical with respect to solvent and solute.

$$\Gamma_2^N = -\Gamma_1^N$$

It is directly related to the Gibbs surface excess.

$$\Gamma_2^N = x_1 \Gamma_2^{(1)}$$

It is calculable from a measured concentration change

$$\Gamma_2^N = -N_o \Delta x_2 / m\Sigma \tag{1a}$$

if the specific area is known, and the solid has a well characterized surface. Its definition gives the relationship between the composite, Γ_2^N, and individual isotherms, n_1^s and n_2^s.

$$\Gamma_2^N = (1/m\Sigma)(n_2^s - x_2(n_1^s + n_2^s)) \tag{1b}$$

The extent of the surface-containing region is not specified by the definition of the surface excess. This region must contain all of the solution whose concentration differs from that of the bulk liquid and may contain any amount of bulk solution. The bulk solution makes no contribution to the value of the surface excess. No assumption is involved in the definition of surface excess as to whether either component forms a monolayer on the surface.

The above definition of the symmetric surface excess and the classical Guoy-Chapman model of the diffuse double layer are combined to show that the surface excess cannot be considered a surface concentration in the presence of an ionized monolayer on an impenetrable solid/liquid interface.

It is postulated that one of the ions of the adsorbed 1:1 electrolyte is surface active and that it forms an ionized monolayer at the solid/liquid interface. All counterions are assumed located in the diffuse double layer (no specific adsorption). Similions are negatively adsorbed in the diffuse double layer. Since the surface-containing region must be electrically neutral, the total moles of electrolyte adsorbed, n_2^s, equals the total moles of counterions in the diffuse double layer which must be equal to the sum of the moles of similions in the diffuse double layer and the charged surface, $A\sigma_0$. These conditions are expressed in Equations 2a,b.

$$n_2^s = (1/N) \int_0^{V^s} v_c dv \tag{2a}$$

$$n_2^s = (1/N) \int_0^{V^s} v_s dv + A\sigma_0 \tag{2b}$$

It is a basic assumption that the preferential adsorption of ions of one sign is the sole source of the electric potential difference between the surface and the bulk solution. The counterions and similions not actually on the surface are assumed to be distributed according to the Boltzmann law, Equation 3, which with Equations 2a and 2b yields Equation 4. The infinite flat plate case is treated first. The volume element, dv, has been replaced by Adl and the exponentials have been combined to give the cosh u term. The factor 10^4 permits the area to be in square meters, consistent with the units of surface excess.

$$v_i = v_b \exp(z_i e\psi/kT) \tag{3}$$

$$n_2^s = A\sigma_0/2 + (10^4 v_b A/N) \cdot \int_{u_0}^{u(L)} \cosh u (dl/du) du \tag{4}$$

The decay of the electric field with distance from the charged surface is obtained from the first integration of the Poisson-Boltzmann equation (20).

$$(du/dl) = -(32\pi e^2 v_b/\epsilon kT)^{1/2} \sinh(u/2) \tag{5}$$

Substitution of 5 into 4 and integration results in an exact expression for n_2^s, Equation 6, which lacks the explicit introduction of the limits of integration.

$$n_2^s/A = (10^4/Ne)(v_b kT/8\pi)^{1/2} (2\sinh(u_0/2) -$$

$$\ln \tanh(u/4) - \cosh(u/2)) \Big|_{u_0}^{u(L)} \tag{6}$$

One of the advantages of using the symmetrical surface excess is that the thickness of the surface region is not at all critical so long as it is allowed to be large enough that the ion concentrations at the limit of the surface region agree with the bulk ion concentrations to the precision desired of the calculation. The value of L was chosen such that $u(L) = 10^{-6}u_0$. Solution concentrations are within 10^{-4} percent of their bulk value with this choice. For the infinite plate case, L varies from about 1400 A. at $C^b = 10^{-3}M$ to about 200 A. at $C^b = 5 \times 10^{-2}M$.

In order to calculate the surface excess (which is a composite isotherm) from the derived individual isotherm for component two, Equation 6, a further assumption is necessary. That chosen is that the partial molecular volumes of the two components in the surface-containing region are the same as in the bulk solution, even though some of the ions are attached to the surface. The error introduced by this assumption leads to a volume of the surface region which might be somewhat too large, but the primary effect is in the second term in Equation 8b, which in these calculations is always less than 10^{-3} times the value of the first term. The error in the calculated surface coverage at saturation is probably less than 1%.

The remaining pertinent equations are Equation 7a for the volume of the surface region and Equation 7b for the equilibrium bulk concentration. Introducing the bulk mole fraction, x_2, and rearrangement yields the relations, Equations 8a and 8b, used with Equations 1a and 1b for numerical calculations.

$$V^s = (n_1{}^s \bar{v}_1 + n_2{}^s \bar{v}_2)(N/10^6) \tag{7a}$$

$$C^b = (10^3/N)(n_2{}^b/(n_1{}^b \bar{v}_1 + n_2{}^b \bar{v}_2)) \tag{7b}$$

$$x_2 = (10^3/N\bar{v}_1 C^b - (\bar{v}_2 - \bar{v}_1)/\bar{v}_1)^{-1} \tag{8a}$$

$$n_1{}^s = (10^6 V^s/N\bar{v}_1) - (\bar{v}_2/\bar{v}_1) n_2{}^s \tag{8b}$$

For use in these equations, $n_2{}^s$ is obtained from Equation 6. The volume of a sodium beta-naphthalenesulfonate molecule, calculated from bond lengths and appropriate van der Waals radii, is taken to be 330 A.^{3}. An average molecular volume of water of 30 A.^{3} was calculated from the density of water at 25.0°C. Most of the numerical work was done on a Honeywell 800 digital computer. The symmetric surface excess and the surface charge densities were calculated over a wide range of surface potentials and concentrations.

Similar calculations, based on the same principles, were carried out for spherical particles. Since the Poisson-Boltzmann equation cannot be integrated analytically in spherical symmetry, a numerical integration was performed. The computer-generated numerical tables of reduced potential as a function of reduced distance of Loeb, Wiersema, and

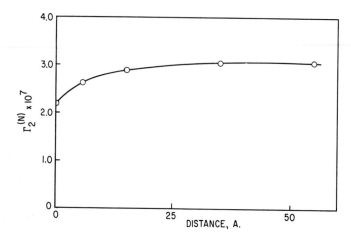

Figure 2. Calculated dependence of the surface excess on distance from an infinite flat solid-liquid interface. $u_o = 4.0$, $C^b = 0.01$, $\sigma_o = 4.41 \times 10^{-7}$

Overbeek (11) were used. The integral over the spherical surface region corresponding to that for an infinite flat plate given in Equation 4 was done numerically using Simpson's rule and results in Equation 9a.

$$n_2{}^s/A = \sigma_o/2 + (10^6 v_b \Delta r/6Na^2)$$

$$\left(f_1 r_1{}^2 + 4 \sum_{i=1}^{(n-1)/2} f_{2i} r_{2i}{}^2 + 2 \sum_{i=2}^{(n-1)/2} f_{2i-1} + f_n r_n{}^2 \right) \quad (9a)$$

$$f_j = \exp(u_j) + \exp(-u_j)$$

It is convenient to redefine the pertinent variables in the terms used by Loeb, Wiersema, and Overbeek so that the available tabular quantities can be used directly as input to the computer. These terms are $q_o = \varkappa a$, $x = 1/\varkappa r$, and $\sigma_o = (\epsilon kT/4\pi e) I(q_o, u_o)$. These substitutions give rise to Equation 9b.

$$n_2{}^s/A = (10^2 \epsilon kT/8\pi e^2 aN)$$

$$\left(q_o I(q_o, u_o) - (\Delta x/6 q_o) \left(f_1 x_1{}^{-4} + 4 \sum_{i=1}^{(n-1)/2} f_{2i} x_{2i}{}^{-4} \right.\right.$$

$$\left.\left. + 2 \sum_{i=2}^{(n-1)/2} f_{2i-1} x_{2i-1}{}^{-4} + f_r x_n{}^{-4} \right) \right) \quad (9b)$$

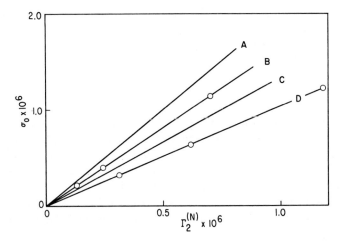

Figure 3. Calculated dependence of the surface charge density on the surface excess. (A) sphere, $u_o \ll 1$, (B) sphere, $u_o = 1.0$, (C) infinite flat plate, $u_o = 7.0$, (D) sphere, $u_o = 7.0$

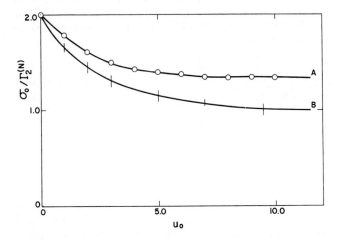

Figure 4. Calculated dependence of the ratio of the surface charge density to the surface excess on the electrostatic surface potential. Curve A is for the infinite flat plate case. Curve B is for a sphere with a 1000 A. radius

The calculated surface excesses for spheres of 1000A. radius are higher than those for a flat plate of the same potential and surfactant concentration by an amount which increases with increasing potential. Typical results of these calculations are included in Figures 3, 4, and 5.

The integration of the spherical analog of Equation 4 can be done analytically only in the Debye-Hückel approximation, $u_o \ll 1$, and leads to Equations 10a and 10b.

$$n_2{}^s = A\sigma_o/2 + A(10^6 \nu_b/3a^2 N)(r^3 - a^3) \qquad (10a)$$

$$\sigma_o \approx 2\Gamma_2{}^N \qquad (10b)$$

The latter equation shows that in the extreme case of low potential, low adsorption, and low ionic strength the negative adsorption of similions in the diffuse double layer region is just half of the positive adsorption due to the charged surface. At the opposite extreme is the high ionic strength region, in which, for spherical particles, $\Gamma_2{}^N = \sigma_o$. In the absence of added salt (the only case considered here) as the ionic strength varies so does the relationship between the surface excess and the charge density of the ionized monolayer.

Experimental

Eastman Kodak technical grade sodium β-naphthalenesulfonate was decolorized with activated charcoal and recrystallized twice from water. Triple A distilled water was used without further distillation. It has a specific conductivity of 3.0×10^{-6} ohm^{-1} cm.$^{-1}$ 0.01 p.p.m. Ca^{2+} and 0.5 p.p.m. other polyvalent cations. The carbon was a special research sample of graphitized Sterling FT supplied by the Cabot Corporation. It was extracted with toluene until the extract showed no detectable difference in its absorption spectrum from that of pure toluene. It was then washed with dilute HCl, extracted with water until the extract showed no chloride, air dried at 90°C. and stored over P$_2$O$_5$. Dispersion of 2.0 grams of this carbon in 10.0 ml. of water gave an increase in pH of 0.9 units to pH 7.4. The area of the carbon was determined from adsorption of nitrogen at 78°K., using 21 A.^{2} per N$_2$ (16, 17). This area, if available in solution, is 17.5 square meters per gram. The corresponding surface average radius is less than 1000 A.

Sodium β-naphthalenesulfonate was chosen as the surface-active electrolyte because its structure is simple and rigid. It does not form micelles, so there is no question as to the species adsorbed on the surface. It is a strong electrolyte and is expected to be essentially completely ionized at saturation coverage. SNS stabilized dispersions flocculate over periods of minutes to months depending on the concentration of SNS. Sterling FTG has a non-polar, non-ionic, hydrophobic surface. The ultimate particles have large, flat, polyhedral surfaces. The particle size distribution of the dry carbon is narrower than that of most colloidal carbons (2).

Two to five grams of carbon were dispersed ultrasonically in 10 ml. of an aqueous solution of sodium β-naphthalenesulfonate using a one gallon Delta-Sonics cleaning tank with a 100 watt, 45 kc. generator for 90 minutes. The dispersion was equilibrated for 24 hours at 25.0°C., followed by centrifugation in a Servall high speed centrifuge. The

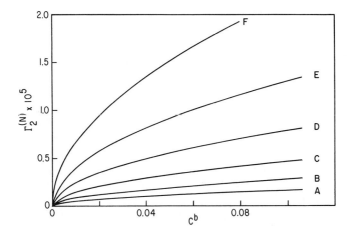

Figure 5. Constant potential isotherms for the adsorption of a 1:1 electrolyte on an infinite flat plate. (A) $u_o = 5.0$ (B) $u_o = 6.0$ (C) $u_o = 7.0$ (D) $u_o = 8.0$ (E) $u_o = 9.0$ (F) $u_o = 10.0$

equilibrium bulk concentration of electrolyte was determined by ultraviolet absorption in a double beam Cary recording spectrophotometer.

The adsorption isotherm was calculated from the measured concentration change. The number of points and their precision suggests that the adsorption values are good to 5%, except at the very lowest concentrations. The absolute accuracy depends on the cleanliness of the carbon surface, which could contain chemisorbed oxygen, and on the completeness of the dispersion process. These possible errors would lead to low values for the experimental surface excess. Comparison of the area per adsorbed ion at apparent surface saturation with the calculated area in different orientations suggests that the entire B.E.T. area is available for adsorption in the dispersions.

Results and Discussion

Values of the surface potential and surface charge density on the carbon black resulting from adsorption of naphthalenesulfonate were calculated from the experimental adsorption isotherm (*see* Figure 7) by means of the derived constant potential isotherms. Both calculated and experimental results are presented in Figures 2 to 8.

Figure 2 shows that, for an infinite flat plate for which u_o is 4.0 and C^b is 0.01M, the surface excess (calculated) is independent of the choice of depth of the surface region provided that the latter is equal to or greater than the Debye "double layer thickness" (33 A. for this concentration). Note that the corresponding surface charge density is larger than the surface excess by a factor of 1.44. That expulsion of similions from the immediate vicinity of the surface is the basic source of the

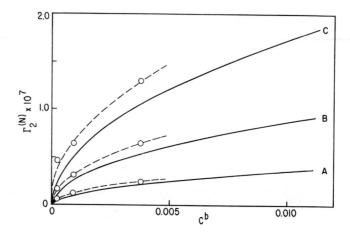

Figure 6. Comparison of constant potential isotherms between the infinite flat plate case, solid line, and that of a spherical particle, dotted line, with a 1000 A. radius. (A) $u_o = 1.0$ (B) $u_o = 2.0$ (C) $u_o = 3.0$

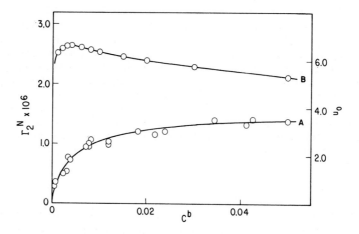

Figure 7. Experimental adsorption isotherm of sodium beta-naphthalenesulfonate on graphitized Sterling FT, A, and the corresponding calculated potential isotherm, B

difference is supported by the observation that this effect is greatest for Debye-Hückel approximation in which the decay of the potential with distance appears slow, and becomes negligible as the surface potential approaches 10 kT/e. If there were no negative adsorption of similions near the surface, the surface charge density would equal the surface excess, analogous to a monolayer of charges. This is the same as the prediction that the ratio of the negative adsorption to the surface charge density depends on ψ_o but is independent of ionic strength (19). Figure 3 shows that the relation between surface charge density and surface excess (both calculated) is linear but that the slope decreases from a high of 2.0 in the Debye-Hückel approximation to a value of 1.35 for a flat plate with a u_o of 7.0 and 1.04 for a sphere of radius 1000 A. and u_o equal to 7.0 Figure 4 shows that the effect of negative adsorption depends on the curvature of the particle as well as on the surface potential; the ratio of σ_o/Γ_2^N falls from the Debye-Hückel limiting value of 2.0 to 1.35 for an infinite flat plate at high potential and to close to 1.0 for spherical colloid particles at high potential. The ratio decreases more rapidly with increasing potential for small spheres than for large spheres.

Figures 5 and 6 show how the calculated values of the surface excess vary with electrolyte concentration for various constant values of the reduced surface potential. These constant potential isotherms are the principal means for interpretation of the experimental adsorption isotherms. Each pair of values of Γ_2^N and C^b corresponds to a specific u_o which can be obtained from these graphs by interpolation. A comparison of the general shape of the constant potential isotherms with the shapes of observed adsorption isotherms of 1:1 surfactants (generally rectangular hyperboloid in the absence of cooperativity) at the graphite/water interface shows that u_o must rise rapidly with increasing surfactant concentration, pass through a maximum and then decrease as the surface region becomes saturated and the ionic strength continues to increase. Zeta-potential measurements on SDS and TMAB stabilized graphon and n-decane dispersions in the absence of added salt (1, 7) fail to confirm this general prediction. A possible difficulty, which has not been resolved is that ζ-potentials calculated from electrokinetic data cannot yet be accurately interpreted.

A given adsorption gives rise to a lower potential on a spherical particle than on a flat plate. The difference is significant, but the limited range of concentrations for which data for spherical particles are available limit their use for quantitative analysis of the experimental adsorption isotherm. This isotherm is given as the lower curve of Figure 7 with the corresponding values of the surface potential calculated from the equations for a flat plate as the upper curve. The maximum corresponds to a surface potential of 170 millivolts and the limiting value at higher concentrations is about 140 millivolts. The experimental adsorption isotherm is not sufficiently precise to determine the exact form of the curve of u_o as a function of concentration as it approaches zero, but the existence of the maximum and its location at 0.004M electrolyte concentration is clearly defined. The surface charge densities for these concentrations and surface potentials are equal to 1.35 Γ_2^N.

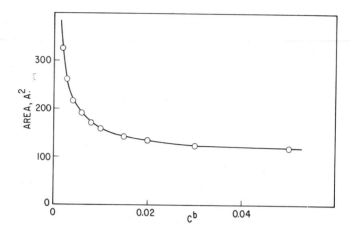

Figure 8. Calculated average surface area per adsorbed naphthalene sulfonate ion

The surface charge density divided by the surface potential gives the integral double layer capacity per unit area of surface. This quantity appears to increase from zero at zero concentration of adsorbable ions to a limiting value of about 110 microfarads per square meter. The differential capacity, obtained as the slope of the curve of surface charge vs. surface potential, poses a problem. It rises normally in very dilute solution but passes through a discontinuity and then becomes negative. A more refined calculation, including both the presence of a Stern layer and the finite size of the adsorbed ions seems necessary for the interpretation of this quantity. The situation seems similar to that on AgI in which a decrease in capacity at high potentials is attributed to the limited number of adsorption sites (14).

Figure 8 gives the average area available to an adsorbed naphthalenesulfonate ion, calculated from the adsorption isotherm and corrected for the difference between the experimental surface excess and the surface charge density. The density of ions "on" the surface is assumed to correspond to σ_0. Sodium naphthalenesulfonate shows no evidence of association or micelle formation (6). Considerable interest lies in the limiting area being only 120 A.^{2} per adsorbed ion. Calculation for a nonhydrated ion with the ring system parallel to the carbon gives a geometrical area of 64 A.^{2}. The ion with two waters of hydration would occupy areas, assuming rectangular close packing, of 136 A.^{2} flat and 78 A.^{2} on end on the surface. It seems likely that the limiting area corresponds to a monolayer of ions with their ring system parallel to the surface. The limiting area cannot be justifiably ascribed to mutual repulsion of the charges since adsorbed paraffin chain sulfonates whose van der

Waals interaction between chains is unlikely to be larger than that between the pi electron systems of the naphthalene rings occupy considerably smaller limiting areas (21).

It is generally accepted that the Guoy-Chapman theory applies to the diffuse portion of the double layer when the reference potential is ψ_d, the potential at the outer Helmholtz plane, and distances are measured with respect to this plane. Negative adsorption of similions is a phenomenon of the diffuse double layer. The concentration of similions in the Stern region is expected to be effectively zero except at the lowest potentials, and the presence of a thin layer of counterions and oriented water will only cause the similion concentration to be actually zero in this thin region. This is not a significant difference. The difference between the surface excess and the surface charge density, being dependent on negative adsorption of similions, is not affected by questions as to the orientation of the adsorbed surfactant ions or the presence of a Stern layer.

The predominant counterion is sodium. It has been shown that sodium is not specifically adsorbed on Hg at even the largest accessible negative potentials (3) and that its specific adsorption potential at the air/solution interface with octadecyl sulfate films is less than kT/e (10). Furthermore, the calculated value of the specific adsorption potential is highly dependent upon the number of possible adsorption sites chosen (15). Trace quantities of Ca^{2+} and other polyvalent impurities (a maximum of two equivalents per 10^6 A.^{2}) will have the effect of reducing the surface potential and electrostatic work of adsorption and increasing the extent of adsorption of surfactant anion. The dramatic effect of Ca^{2+} on sodium dodecyl sulfate and sodium dodecylbenzene sulfate stabilized graphon dispersions (23) involved concentrations of Ca^{2+} more than hundredfold larger than could possibly have been present in this work. It is in the low concentration range of the adsorption isotherm that the sensitivity of adsorption to the presence of polyvalent cations will be the greatest. In view of the very small amount of impurities present, it seems safe to conclude that they do not significantly influence these experimental results.

Summarizing briefly, the major conceptual point of this paper is the prediction that the thermodynamic surface excess and the electrostatic surface charge density are not equal. This result follows directly from the negative adsorption of similions in the diffuse double layer. For strongly adsorbed surfactants on spherical particles, the surface potential will generally be so high that σ_o and Γ_2^N can be equated. However, for large particles with flat surfaces, the ratio σ_o/Γ_2^N does not decline below about 1.33 for potentials as high as 10 kT/e. Limiting areas of adsorbed ions calculated by equating σ_0 and Γ_2^N will be too high. For lower potentials, regardless of the surface curvature, the effect will be even higher.

Acknowledgments

The authors are grateful to A. I. Medalia and the Cabot Corporation for the special preparation of the carbon black, to Leon Dormant of the University of Southern California for determination of its B.E.T. area, and to the U.S.C. computer Sciences Laboratory for the use of the Honeywell 800.

Literature Cited

(1) Anderson, P. V., *Trans. Faraday Soc.* **55**, 1421 (1959).
(2) Cabot Corporation, "Cabot Carbon Blacks Under the Electron Microscope," 2nd ed., Vol. 6, no. 12, p. 87 (1953).
(3) Grahame, D. C., *Chem. Rev.* **41**, 441 (1947).
(4) Guggenheim, E. A., Adam, N. K., *Proc. Royal Soc.* **A139**, 218 (1933).
(5) Guoy, G., *Ann. Phys.* **7**, 129 (1917).
    Chapman, P. L., *Phil. Mag.* **25**, 475 (1913).
(6) Hattori, K., *Kao Soap Co. Ltd. publication*, Wakayama, Japan (1964).
(7) Haydon, D. A., Phillips, J. N., *Trans. Faraday Soc.* **54**, 698 (1958).
(8) Kipling, J. J., "Adsorption from Solutions of Non-Electrolytes," Academic Press, New York, 1965.
(9) Kruyt, H. R., ed., "Colloid Science," Vol. I, Elsevier, New York, 1952.
(10) Levine, S., Mingins, J., Bell, G. M., *J. Phys. Chem.* **67**, 2095 (1963).
(11) Loeb, A. L., Wiersema, P. H., Overbeek, J. Th. G., "The Electrical Double Layer Around a Spherical Colloid Particle," Massachusetts Institute of Technology Press, Cambridge, 1961.
(12) Lyklema, J., *Trans. Faraday Soc.* **59**, 418 (1963).
(13) Lyklema, J., *Discussions Faraday Soc.* **42**, 81 (1967).
(14) Lyklema, J., Overbeek, J. Th. G., *J. Colloid Sci.* **16**, 595 (1961).
(15) Mingins, J., Pethica, B. A., *Trans. Faraday Soc.* **59**, 1892 (1963).
(16) Pierce, C., Ewing, B., *J. Phys. Chem.* **68**, 2562 (1964).
(17) Pierce, C., Ewing, B., *J. Am. Chem. Soc.* **84**, 4070 (1962).
(18) van den Hull, H. J., Lyklema, J., *J. Colloid and Interface Sci.* **23**, 500 (1967).
(19) van Olphen, H., "Introduction to Clay Colloid Chemistry," Interscience Publishers, New York, 1963.
(20) Vervey, E. J. W., Overbeek, J. Th. G., "Theory of the Stability of Lyophobic Colloids," Elsevier, Amsterdam, 1948.
(21) Vold, R. D., Sivaramakrishnan, N. H., *J. Phys. Chem.* **62**, 984 (1958).
(22) Zettlemoyer, A. C., *Proc. 155th Natl. A.C.S. Meeting, San Francisco, 1968.* Kendall Award Address. MOO8.
(23) Zettlemoyer, A. C., Skewis, J. D., Chessick, J. J., *J. Am. Oil Chemists' Soc.* **39**, 280 (1962).

RECEIVED October 26, 1967.

13

The Effect of Hydrocarbon Chain Length on the Adsorption of Sulfonates at the Solid/Water Interface

T. WAKAMATSU and D. W. FUERSTENAU

College of Engineering, University of California, Berkeley, Calif.

> *Isotherms for the adsorption at the alumina-water interface of sodium alkylsulfonates containing 8, 10, 12, 14, and 16 carbon atoms were determined at constant pH, temperature, and ionic strength. Electrophoretic mobilities were measured for the same conditions. It is clearly shown that the adsorption isotherm for such detergents consists of three distinct regions. In Region 1 where adsorption occurs by ion exchange with chloride ions used to control ionic strength, the isotherms are approximately characterized by the same line. The onset of the increased adsorption owing to association of adsorbed detergent ions denoting Region 2 depends on the hydrocarbon chain length. In Region 3, the electrokinetic potential is reversed and the isotherms exhibit a decreased dependence on concentration because of electrostatic repulsion at the surface.*

The adsorption of ionic surfactants at solid-water interfaces is of great technological importance in such diverse fields as water renovation, detergency, mineral flotation, and corrosion inhibition. The complex nature of the adsorption of surfactants at the solid-water interface is controlled by the nature of the adsorbing species itself, the properties of the solid adsorbent, and the composition of the aqueous solution. Dependent on these factors, ionic surfactant-solid systems can be classified into three broad types, namely (1) those in which the surfactant adsorbs as counterions in the double layer through coulombic interaction with the charged surface, (2) those in which the surfactant adsorbs by covalent bond formation with the solid surface, and (3) those in which the surfactant adsorbs through hydrophobic bonding of the hydrocarbon chain to a nonpolar solid such as graphite. In all such adsorption processes, the hydrocarbon chain of the surfactant plays a dominant role in the behavior of the system. The importance of the hydrocarbon chain in aqueous interfacial phenomena has been demonstrated by a variety of experimental methods, for example, by electrokinetic studies (*11*), flotation behavior (*2*), adsorption measurements (*12*), and coagulation behavior (*8*).

The adsorption of ionic surfactants by nonpolar solids such as graphite would be expected to depend markedly on the length and configuration of the hydrocarbon chain since it is through the hydrocarbon chain that the surfactant bonds to the solid. The investigation of Skewis and Zettlemoyer (*9*) is quite typical of the kinds of effects obtained in such adsorption systems. One of the few investigations reported in the literature on the adsorption of ionic surfactants of varying chain length on charged solids in aqueous media is that of Jaycock, Ottewill, and Rastogi (*5*). Using colloidal silver iodide and aqueous solutions of pyridinium bromides of various chain lengths, they found that at low coverages, the adsorption density was approximately the same for each of the surfactants but that the adsorption increased catastrophically at some critical concentration dependent on chain length. The use of silver iodide as the adsorbent for such studies complicates the situation considerably because, as is now well known, silver iodide is partially hydrophobic (*4, 13*). Thus, the adsorption in this system must be some combination of coulombic interaction and hydrophobic bonding with the surface. Essentially the only other investigation of the adsorption of ionic surfactants of various chain lengths at mineral-water interfaces is that of Tamamushi and Tamaki (*12*), who determined the adsorption of alkylammonium chlorides on alumina. They attempted to explain their isotherms in terms of a Brunauer-Emmett-Teller (B.E.T.) type of equation, but the validity of their approach is questionable because of their neglect of electrical effects.

Recently (*10*), it was demonstrated that the adsorption of alkylsulfonates at the alumina-water interface is a system in which the surfactant adsorbs as counterions in the electrical double layer. This work showed that the isotherm for the adsorption of sodium dodecyl sulfonate at the alumina-water interface is characterized by three distinct regions: Region 1 in which the detergent ions adsorb individually through coulombic attraction for the surface; Region 2 in which the adsorption is enhanced through association of the hydrocarbon chains of the adsorbed surfactant ions; and Region 3 in which the charge in the Stern plane exceeds the surface charge with the resulting electrostatic repulsion acting to retard adsorption. In the experimental investigation discussed in the present paper, details of the role of the hydrocarbon chain in the adsorption process at the alumina-water interface have been studied for a series of alkyl sulfonates with varying chain lengths. In this work, pH, ionic strength, and temperature were maintained constant. In particular, our purpose has been to determine how each of these adsorption regions depends on the hydrocarbon chain of the surfactant.

Experimental

Materials and Methods. For the solid adsorbent, α-alumina (Linde "A") of 99.95% purity was used. Its specific surface area, as measured by krypton gas adsorption and by stearic acid adsorption from benzene was found to be 15 meter^2/gram. Its zero-point-of-charge occurs at pH 9.1. Details about the characterization of this material have been described in a previous paper (15). The alkylsulfonates were prepared by neutralizing high purity sulfonic acids (C8, C10, C12, C14, and C16) with sodium hydroxide and by recrystallizing the sodium salt from hot absolute ethyl alcohol. Detailed infrared spectroscopic analysis of these reagents showed them to be pure sulfonates with but a trace amount of water as the only impurity. Only the C14 reagent appears not to be completely free of shorter chain homologs. All inorganic chemicals were reagent grade. Conductivity water prepared in a quartz still was used for all solutions.

The adsorption experiments were carried out in a 600 ml. beaker that was stoppered with a Teflon disc containing holes for a thermometer, gas inlet and outlet, and pH electrodes. The cell containing the alumina and the surfactant solution was maintained at a constant temperature by immersing the cell in a thermostatically controlled water bath. An atmosphere of purified nitrogen was maintained over the cell. The experimental conditions were held constant at pH 7.2, 25°C., and $2 \times 10^{-3} M$ ionic strength (adjusted with sodium chloride). The system was stirred by means of a magnetic, Teflon-covered stirring bar for four hours. About two hours was required for adjusting pH, and the remaining time was used for attaining equilibrium and for sampling. The method used for the determination of sulfonate concentration is the well-known methylene blue complex method (6, 7).

Electrophoretic mobilities of the alumina particles were determined for the same conditions as were used to obtain the adsorption isotherms. For this purpose, a sample of the alumina suspension was transferred to the electrophoresis cell for measurement of the electrophoretic mobilities. A Zeta-Meter was used for this part of the program.

Results. The isotherm for the adsorption of sodium dodecyl sulfonate by alumina at constant pH (pH 7.2) and ionic strength ($2 \times 10^{-3} M$) is given in Figure 1 to illustrate specifically the nature of the isotherm obtained for the adsorption of a detergent from aqueous solution. In this figure, the amount of sulfonate adsorbed per unit area of alumina is plotted logarithmically as a function of the equilibrium concentration of alkylsulfonate in solution. The adsorption isotherm consists of three distinct regions: Region 1, which is characterized by a low increase in adsorption with increasing surfactant concentration; Region 2, by an

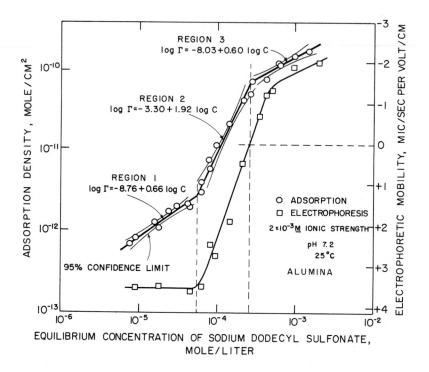

Figure 1. The electrophoretic behavior and isotherm for the adsorption of sodium dodecyl sulfonate from aqueous solution at pH 7.2, 25°C., and 2×10^{-3}M ionic strength (NaCl). The 95% confidence limits for the three straight-line regions of the adsorption isotherm are shown

abrupt increase in the slope of the isotherm; and Region 3, again by a decreased dependence of adsorption on sulfonate concentration. To correlate changes in the adsorption process with the isotherm, the electrophoretic behavior of alumina in the presence of sodium dodecyl sulfonate is also included in Figure 1. In Figure 2, isotherms for the adsorption of C8 to C16 sulfonates are presented to show how the chain length of the detergent affects the adsorption process.

The electrophoretic mobilities of alumina for the same conditions as used for determination of the isotherms are presented in Figure 3. This figure, together with Figures 1 and 2, shows that the mobility is positive in sign in Regions 1 and 2 but is negative in Region 3. Region 1 is characterized by the electrophoretic mobility being nearly independent of the sulfonate concentration. The transition between Region 1 and Region 2 is marked by a sharp change in the electrophoretic mobility-vs.-concentration curve whereas the transition between Regions 2 and 3 occurs at concentrations where the mobility is zero. Clearly, the electrophoretic behavior of the alumina depends markedly on the number of carbon atoms in the hydrocarbon chain of the detergent. The adsorption density marked as a monlayer in Figure 2 is that for a closely packed layer of vertically oriented alkylsulfonate ions.

Discussion of Results

The Adsorption Isotherm. In a previous paper (*10*), it was shown that the isotherm for the adsorption of a detergent at the solid-water interface can be characterized by three distinct regions but at that time the data were not analyzed statistically. Statistical analysis of the data presented in Figure 1 shows that the isotherm corresponding to each of the three regions can be characterized by the following straight lines:

$$\text{Region 1: } \log \Gamma = -8.76 + 0.66 \log C$$
$$\text{Region 2: } \log \Gamma = -3.30 + 1.92 \log C$$
$$\text{Region 3: } \log \Gamma = -8.03 + 0.60 \log C$$

where Γ is the amount of sulfonate adsorbed in mole per cm.^{2} and C is the molar concentration of sulfonate in the bulk solution. Statistical analysis of these data shows that the slope of these three straight lines at the 95% confidence level (*14*) is 0.66 ± 0.19, 1.92 ± 0.25, and 0.60 ± 0.22 for Regions 1, 2, and 3, respectively. Region 1 for sodium dodecyl sulfonate under these experimental conditions applies up to a detergent concentration of $6 \times 10^{-5} M$. The adsorption is characterized by Region 2 between $6 \times 10^{-5} M$ and $3 \times 10^{-4} M$ under these conditions, and by Region 3 above $3 \times 10^{-4} M$ sulfonate.

Further evidence for three distinct modes of adsorption can be seen in the electrophoretic behavior of alumina in the presence of sodium dodecyl sulfonate. Below $6 \times 10^{-5} M$, the electrophoretic mobility is nearly independent of concentration, but at this concentration the slope of the mobility–*vs.*–concentration curve abruptly changes. At $3 \times 10^{-4} M$ dodecyl sulfonate concentration, the electrophoretic mobility reverses its sign, indicating that the charge in the Stern layer now exceeds the surface charge in absolute magnitude.

This experimental evidence clearly shows the complicated nature of the adsorption isotherm for a detergent at the polar solid-aqueous solution interface.

The Effect of Alkyl Chain Length on Adsorption. Figure 2 shows that the adsorption isotherms all have somewhat the same general characteristics only the concentrations at which the effects occur appear to depend on the alkyl chain length. Consequently, the adsorption behavior in each of the three regions of the isotherms will be discussed separately and will be interpreted in terms of the role that the hydrocarbon chain plays in the adsorption process.

It has already been established (*10*) that alkylsulfonate ions adsorb at the positively charged alumina-water interface as counterions in the electrical double layer. Chemisorption is absent. The electrical double

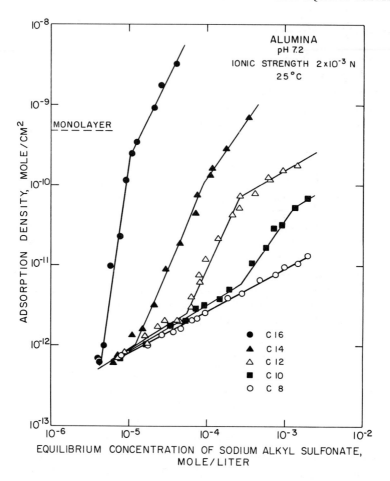

Figure 2. Adsorption isotherms for sodium alkylsulfonates of different hydrocarbon chain lengths on alumina at pH 7.2, 25°C., and 2 × 10^{-3}M ionic strength

layer model will then be used to interpret the adsorption behavior observed in these experiments.

In Region 1, alkylsulfonate ions are considered to adsorb as individual counterions in competition with the chloride ions used to control the ionic strength. If the adsorption is of non-associated sulfonate ions by ideal exchange in the diffuse layer, a single line independent of chain length should characterize the adsorption in Region 1. For the C16 sulfonate, at the lowest concentration which could be studied, the adsorption already has exceeded that of Region 1. On the other hand, Figures 2 and 3 show that the adsorption of C8 sulfonate is restricted entirely to Region 1. For C8, C10, C12, and C14 this adsorption is characterized approximately by a single line, whose slope is about 0.7 instead of the expected slope of 1.0. Such deviations from ideal exchange are not yet clearly understood but are considered to reflect differences in the ability of surfactant ions compared with chloride ions to penetrate the diffuse layer region. This is a subject of continued research.

Region 2 is characterized by a marked change in the slope of the adsorption isotherms. This results from the onset of association of the hydrocarbon chains of the surfactant ions adsorbed in the Stern plane. The mean separation distance of adsorbed ions under these conditions is about 70 A., which approximates the mean separation distance in bulk at the c.m.c. In such adsorption phenomena, there is a relationship between this asociation and the formation of micelles in bulk solution. For example, electrokinetic studies (*1*) on quartz at neutral pH showed that alkylammonium ions associate in the Stern plane when their bulk concentration is approximately one hundredth of the c.m.c. This association which has been called hemimicelle formation (*3*), gives rise to a specific adsorption potential which causes the adsorption to increase markedly and brings about a reversal in the sign of the potential at the Stern plane. The hemimicelle concentration, that is the bulk concentration necessary

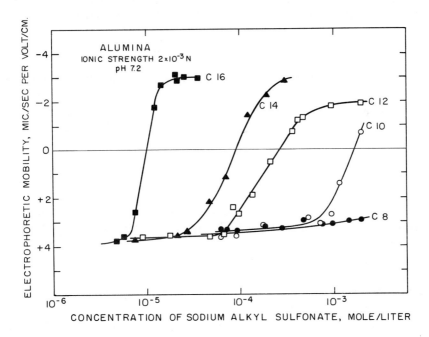

Figure 3. The electrophoretic mobility of alumina at pH 7.2 and 2 × 10⁻³M ionic strength as a function of the concentration of sulfonates with various hydrocarbon chain lengths

to induce association in the interface, increases with decreasing chain length. At pH 7.2 on alumina this concentration is $4 \times 10^{-6}M$ for C16, $1.4 \times 10^{-5}M$ for C14, $6 \times 10^{-5}M$ for C12, $7 \times 10^{-4}M$ for C10, and is above $2 \times 10^{-3}M$ for C8 sulfonate. The slopes of the isotherms are 0.56 for C8, 1.33 for C10, 1.92 for C12, 2.50 for C14, and 5.80 for C16 sulfonate. Thus, with increasing chain length, the competition between RSO_3^- and Cl^- for sites at the surface strongly favors the adsorption of the detergent.

In a previous publication (10) it was shown that adsorption in Region 2 where association of the hydrocarbon chains is leading to extensive adsorption and eventual reversal of the sign of the mobility— i.e., ψ_δ—is given by

$$\Gamma_\delta = 2rC \exp\left[\frac{-zF\psi_\delta - n\phi}{RT}\right] \qquad (1)$$

where Γ_δ is the adsorption density in the Stern plane in moles/cm.^{2}, r is the effective radius of the adsorbed ion, C is the bulk concentration in moles/cc., z is the valence of the adsorbed ion, F is Faraday's constant, ψ_δ is the potential in the Stern plane, ϕ is the cohesive energy per mole of CH_2 groups, n is the effective number of associating CH_2 groups per hydrocarbon chain, R is the gas constant, and T the absolute temperature. Thus, expressing adsorption on a logarithmic basis, we have

$$\frac{d\ln\Gamma}{d\ln C} = 1 - \frac{zF}{RT}\frac{d\psi_\delta}{d\ln C} - \frac{\phi}{RT}\frac{dn}{d\ln C} \qquad (2)$$

The variation of n with concentration expresses the fact that in Region 2 complete removal of each CH_2 group of the surfactant from water is only possible at a monolayer. In bulk systems the analogous processes are the pre-association into dimers, trimers, etc. just below the critical micelle concentration.

From the slopes of the adsorption isotherms and the mobility-concentration curves, it is possible to evaluate with Equation 2, $dn/d\ln C$— i.e., the term which expresses the effective number of CH_2 groups removed from the aqueous environment. By this means, the values of $dn/d\ln C$ are found to be 9 for C16, 4 for C14, 2 for C12, 2 for C10, and 0 for C8. A comparison can be made for the C16 sulfonate by considering the mobility curves and adsorption isotherms. Observation of the electrophoretic mobility curve for the C16 surfactant shows that hemimicelle formation begins at $6 \times 10^{-6}M$ and that the mobility curve again becomes independent of concentration at $1.5 \times 10^{-5}M$, this latter concentration coinciding with monolayer coverage in the expected manner. Thus, n has an effective value of zero at about $6 \times 10^{-6}M$ and a value of 15 at monolayer coverage (assuming that the terminal CH_3 groups are exposed to the solution). This leads to an estimated value of $dn/d\ln C$ of approximately 16. As pointed out previously (*10*), a limitation of this treatment is that throughout Region 2, the fraction of the sulfonate ions that adsorb in the Stern layer increases, reaching unity at the end of Region 2. Since Equation 2 depends on the assumption that the adsorption in the Stern layer is proportional to the total adsorption, the value of $dn/d\ln C$ calculated from the adsorption will be too low.

The end of Region 2 is characterized by the fact that the mobility passes through zero. When the mobility is zero, the Stern-Grahame expression for the adsorption of a specifically adsorbing surfactant ion can be expressed as

$$(\Gamma_\delta)_o = 2rC_o \exp\left(-\frac{N\phi}{RT}\right) \quad (3)$$

where N is the number of carbon atoms in the hydrocarbon chain. Here we must assume that the effective fraction of the CH_2 groups removed from aqueous environment when ψ_δ is zero must be independent of chain length. Putting Equation 3 in logarithmic form and rearranging terms yields:

$$\ln C_o - \ln\frac{(\Gamma_\delta)_o}{2r} = -N\frac{\phi}{RT} \quad (4)$$

The numerical values of $(\Gamma_\delta)_o$ and C_o are tabulated in Table I.

Table I.

	$(\Gamma_\delta)_o$	C_o
C16	1.6×10^{-10} mole/cm.^{2}	1.0×10^{-5} mole/liter
C14	1.1×10^{-10}	9.0×10^{-5}
C12	7.9×10^{-11}	2.8×10^{-4}
C10	6.0×10^{-11}	1.7×10^{-3}

From these numbers, it can be seen that the change of $(\Gamma_\delta)_o$ when increasing the number of CH_2 groups in the alkyl chain length from C10 to C16, is 6.0×10^{-11} to 1.6×10^{-10}, while the variations in the values of $(C_\delta)_o$ is 10^{-5} to 1.7×10^{-3}. We do not know the effective radius of each sulfonate ion at zero mobility, but it certainly must correspond approximately to the thickness of the adsorbed layer. Perhaps it may even increase with the number of carbon atoms in the hydrocarbon chain. However, as the variation in the value of $(C_\delta)_o$ is very large compared with that of $(\Gamma_\delta)_o$, it will be assumed that $\ln(\Gamma_\delta)_o/2r$ is a constant independent of the chain length. Thus, if we plot the logarithm of $(C_\delta)_o$ against N, a straight line should be obtained. Accordingly, in Figure 4, the concentration of sodium alkylsulfonate in solution corresponding to zero mobility is plotted as a function of the number of carbon atoms in the alkyl chain. From the slope of this line, the value of ϕ is calculated to be $-0.95\ RT$, a value in good agreement with previous studies of amine salts on quartz (2, 11).

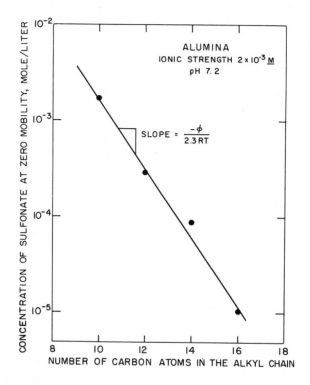

Figure 4. *Variation of the concentration of sulfonate necessary for zero electrophoretic mobility of alumina as a function of the number of carbon atoms in the alkyl chain*

In Region 3 the solpe of the isotherm is lower than in Region 2. Here the surfactant ions probably adsorb by a somewhat different mechanism. As shown in Figure 2, the values of ψ_δ must be negative in this region, and consequently the adsorbed sulfonate ions should be subjected to electrostatic repulsion in the adsorption process. Thus, reducing the slope of the isotherms in this region. In Region 3 the slope of the C16 isotherm is 2.16, that of the C14 is 1.50 and that of the C12 isotherm is 0.60. Although only three values for this slope could be obtained from our studies, it is apparent that the value becomes greater with increasing hydrocarbon chain length. This tendency again can be attributed to the increased attraction between hydrocarbon chains with increase in chain length. Further, the adsorbed detergent may tend to orient differently at the surface because of the electrostatic repulsion upon reversal of ψ_δ.

Summary

As a part of a study on the mechanism of adsorption of alkyl sulfonates at the alumina-water interface, the role of hydrocarbon chain length in the adsorption process has been investigated by adsorption and electrophoresis measurements with sodium alkylsulfonates containing 8, 10, 12, 14, and 16 carbon atoms at constant pH, temperature, and ionic strength. The adsorption isotherms have clearly been shown to consist of three distinct regions, depending upon the intermolecular behavior of the hydrocarbon chain. In Region 1 where the detergent ions adsorb in the double layer in competition with the chloride ions used to control ionic strength, the isotherms are characterized by approximately the same straight line. In Region 2 the adsorbed detergent ions associate, resulting in a sharp increase in the adsorption density. The onset of this association occurs at lower bulk concentrations as the hydrocarbon chain length is increased. In Region 3 the adsorption isotherms again have a decreased slope; in this region the electrokinetic potential is reversed, resulting in electrostatic repulsion between the adsorbed ions. By means of the Stern-Grahame model of the double layer under conditions where the electrophoretic mobility of the alumina is zero, the cohesive free energy per mole of CH_2 groups has been calculated to be $-0.95\ RT$.

Acknowledgments

The authors wish to acknowledge the National Institute of Health (Grant No. WP-00692) for support of this research. Discussions with T. W. Healy are also acknowledged.

Literature Cited

(1) Fuerstenau, D. W., *J. Phys. Chem.* **60,** 981 (1956).
(2) Fuerstenau, D. W., Healy, T. W., Somasundaran, P., *Trans. AIME* **229,** 321 (1964).
(3) Gaudin, A. M., Fuerstenau, D. W., *Trans. AIME* **202,** 958 (1955).
(4) Hall, P. G., Tompkins, F. C., *Trans. Faraday Soc.* **58,** 1734 (1962).
(5) Jaycock, M. J., Ottewill, R. H., Rastogi, M. C., *3rd Intern. Congr. Surface Activity* Vol. II, 283 (1960).
(6) Jones, J. H., *J. Assoc. Agr. Chemists* **28,** 398 (1945).
(7) *Ibid.,* **28,** 409 (1945).
(8) Ottewill, R. R., Rastogi, M. C., *Trans. Faraday Soc.* **56,** 880 (1960).
(9) Skewis, J. D., Zettlemoyer, A. C., *3rd Intern. Congr. Surface Activity* Vol. II, 401 (1960).
(10) Somasundaran, P., Fuerstenau, D. W., *J. Phys. Chem.* **70,** 90 (1966).
(11) Somasundaran, P., Healy, T. W., Fuerstenau, D. W., *J. Phys. Chem.* **68,** 3562 (1964).
(12) Tamamushi, B., Tamaki, K., *Proc. 2nd Intern. Congr. Surface Activity* **3,** 449 (1958).

(13) Tcheurekdjian, N. Zettlemoyer, A. C., Chessick, J. J., *J. Phys. Chem.* **68**, 773 (1964).
(14) Volk, W., "Applied Statistics for Engineers," p. 236, McGraw-Hill, New York, 1958.
(15) Yopps, J. A., Fuerstenau, D. W., *J. Colloid Sci.* **19**, 61 (1964).

RECEIVED November 24, 1967.

14

Adsorption of Dyes and Their Surface Spectra

A. H. HERZ, R. P. DANNER, and G. A. JANUSONIS
Research Laboratories, Eastman Kodak Company, Rochester, New York, 14650

The adsorption of dyes and particularly of cyanines at silver halide, silver or mica substrates is generally accompanied by changes in the dye spectra. Depending on the system, these changes have been quantitatively evaluated either by reflectance measurements with application of the Kubelka-Munk relation or by transmission spectroscopy. Surface concentrations and saturation coverages of the dyes as well as surface areas and average particle dimensions of the substrates were obtained. These values generally agreed well with independently determined measurements and yielded Langmuir adsorption coefficients, apparent standard free energies of adsorption, and probable orientations of the adsorbed dyes.

It is the purpose of this paper to describe methods for determining and interpreting dye spectra in aqueous dispersions of silver halides and other substrates. Such spectra can be utilized for the direct measurement of surface concentrations of dyes from which, in turn, the surface area of the substrate can be derived. The techniques involved are not limited to a specific dye class but will be illustrated in this paper by the behavior of cyanine dyes.

The known factors which influence the solution spectra of cyanines have been recently reviewed (14, 46, 73). It will be sufficient to summarize here some of the concentration-dependent spectral properties for the specific case of 1,1′-diethyl-2,2′-cyanine; this cyanine was employed in many of the present experiments. In alcohol as in dilute water solutions this dye, which will be referred to as Pseudocyanine, although it has also been called Pseudoisocyanine, appears to exist only in an extended configuration and has its maximum absorption near 523 n.m. This transition is associated with isolated molecules and is referred to as the molecular or M-band and is accompanied by a subsidiary vibrational shoulder near 490 n.m. On increasing the dye concentration beyond about $5 \times 10^{-5}M$, the intensity of the molecular band decreases and a new maximum appears near 480 n.m.; this band is associated with formation of a dimer and is accordingly called the D-band. A further increase in dye concentration causes broadening of the dimer band and enhanced absorption in the shorter or hypsochromic wavelength region. With some cyanines, although not with Pseudocyanine, this shift to hypsochromic absorption is accompanied by formation of definable maxima or H-bands. These are believed to arise from the formation of polymers other than the dimer.

These concentration-dependent spectral changes may also be brought about by inert electrolytes, surfactants, or some organic solvents and have been related to dye-dye interactions. An alternate view (42, 43, 44), which ascribed these metachromatic changes to dye-counterion interaction, has been refuted on the basis of counterion-activity determinations (51). Conversely, there seems to be no disagreement about the polymeric association that occurs in aqueous Pseudocyanine at concentrations above $5 \times 10^{-3}M$ (25°C.). In such relatively concentrated solutions, an abnormal increase of viscosity is observed, together with the appearance of a narrow band of high absorptivity which is located at wavelengths longer than the molecular band, at approximately 570 n.m. This new and fluorescent transition, referred to as the J-band, was first identified by Jelley (26, 27) and by Scheibe (56, 57, 58, 59) and is thought to be caused by aggregates of stacked cyanine molecules held by van der Waals forces in a close-packed and possibly helical configuration (39, 40, 53). Absorption within this band is considered to arise from an electronic transition perpendicular to the chromophore of individual molecules and parallel to the axis of the multimolecular array (25, 41, 56, 57, 58, 59).

From this summary it is apparent that water solutions of Pseudocyanine above *ca.* $5 \times 10^{-5}M$ contain a multicomponent mixture of differently absorbing species. Hence, the resulting solution spectra are neither expected nor found to exhibit isosbestic points (12). However, as will be shown, in the presence of appropriate adsorbents the spectra of Pseudocyanine can be drastically modified by imposition of an equilibrium between monomeric dye in solution and its J-aggregate at the substrate surface.

Experimental

Materials. Dye structures are given in the figures. Pseudocyanine (7, 21) (1,1'-diethyl-2,2'-cyanine chloride, $C_{23}H_{23}N_2Cl \times H_2O$) had a molar absorptivity of $7.0 \pm 0.2 \times 10^4 M^{-1}$ cm.$^{-1}$ at 523 n.m. in water

(23°C.) at concentrations below $5 \times 10^{-5}M$. The p-toluenesulfonate (pts) salt of this cyanine was also used; it gave identical surface spectra and adsorption data in silver and silver halide dispersions (23). Astraphloxin (55) (1,1'-diethyl-3,3,3',3'-tetramethylindocarbocyanine pts, $C_{34}H_{40}O_3N_2S \times 0.5\ H_2O$) had a molar absorptivity of $13.8 \pm 0.4 \times 10^4 M^{-1}$ cm.$^{-1}$ at 542 n.m. at concentrations below $5 \times 10^{-5}M$ in water. At higher concentrations both Pseudocyanine and Astraphloxin failed to obey Beer's law, owing to dye-dye interaction leading to formation of new absorption bands (12).

Synthetic mica was obtained from the Minnesota Mining and Manufacturing Co., St. Paul, Minn., as Burnil Micro Plates HX-600. This colorless powder is described by the manufacturer as consisting of platelets 20-100 A. thick and ten times as long. Colloidal silver with an average particle diameter of about 150 A. was prepared by the dextrin-reduction of hydrated silver oxide, according to the general procedure of Carey Lea (67). After addition of gelatin, the dialyzed parent dispersion was adjusted with bromide to pBr 3 and with acid to pH 6.5; it contained 5% gelatin by weight and was $2.5 \times 10^{-1}M$ in respect to silver. On dilution with water, a clear yellow sol resulted; it obeyed Beer's law at least up to $3 \times 10^{-4}M$ and exhibited a molar absorptivity of $1.45 \times 10^4 M^{-1}$ cm.$^{-1}$ at its 405 n.m. absorption maximum. One of the silver bromide dispersions was already used in previous adsorption experiments (*see* References 22 and 23, Dispersion D). It contained 6 mole-percent iodide and had a specific surface area of 1.1 meter^2/gram AgX, as determined from the saturation coverage with Pseudocyanine. A molecular area of 57 A.^{2} had been assigned to this dye. (Adsorption determinations with Pseudocyanines in AgBr dispersions whose surface areas were measured by three independent methods have previously led to a limiting area of 54 ± 4 A.^{2} per dye molecule (23). Further work has supported the upper limit as the more accurate value; accordingly, we have adopted a molecular area of 57 A.^{2} throughout our calculations.) This relatively coarse AgBr dispersion was generally used at a concentration of $7 \times 10^{-2}M$ in 0.2% gelatin at pH 6.5 and pBr 3. The other silver bromide dispersion was a Lippmann type (47) and consisted of small, nearly spherical particles with an average diameter of 585 ± 35 A. to 700 ± 40 A., as estimated from electronmicrographs and light scatter, respectively. The source of this considerable discrepancy was not further investigated and the Lippmann AgBr dispersion was used in a dilute state at $5 \times 10^{-4}M$ or less, at the already cited Br$^-$, H$^+$, and gelatin concentrations. In most of the experiments described, gelatin was present as a convenience. Since gelatin as well as other organic polymers may induce metachromacy in cyanine dyes (1, 2, 6, 11, 28, 29), it was necessary to establish if the gelatin used in the dispersions interfered with the measurements. This was done by carrying out adsorption determinations with Pseudocyanine in silver and silver bromide dispersions both in the absence and presence of gelatin under otherwise identical conditions. In agreement with expectations (9), no evidence was found that gelatin influenced surface saturation by the dye at adsorption equilibrium (62, 70). However, with one particular silver dispersion it was noted that, after saturation coverage of the surface by Pseudocyanine had been attained with concomitant formation of a *J*-band at 580 n.m., excess dye produced a second *J*-band at 570 n.m.

This second band was readily differentiated from the former and control experiments carried out in the absence of silver demonstrated that the 570-band was caused by interaction of the dye with the specific gelatin derivative used in that system.

Procedures. Unless otherwise specified, spectra or adsorption determinations were made after equilibration of dye and substrate for at least 1 hr. at 23° ± 1°C. The coarse AgBr tended to settle and was agitated even in the spectrophotometric cell. The silver halides were always handled under photographic safelights. Conventional adsorption isotherms were obtained by usual centrifugation and phase-separation procedures (23). Differential dye spectra were determined with a double-beam spectrophotometer, usually a Beckman DK-2 Spectroreflectometer, by placing the dyed dispersion in the sample cell (0.1 to 40 mm.) and by using the identical but undyed dispersion in the comparison cell as the spectral reference. Because colloidal silver absorbs strongly in the blue spectral region, it was not feasible to measure differential spectra with this dispersion at wavelengths below ca. 480 n.m. With the weakly scattering and small-particle dispersions of silver, mica, and silver bromide of the Lippmann type, transmission spectrophotometry proved to be practical. However, with coarse and highly turbid silver halide dispersions, the diffuse reflectance, R, was determined under conditions such that R_∞ prevailed—i.e., an increase in cell length had no measurable influence on reflectivity. In the reflectance measurements the undyed dispersion was again used as the spectral reference and experiments with polarizing screens showed that the light reflected from the silver halide dispersions contained no appreciable Fresnel components.

Results

The transmission spectra of Figure 1A show that, with progressive additions of synthetic mica to a constant concentration of Pseudocyanine solution, the intensity of the M-band diminished, with the concomitant appearance of new maxima. The first new peak appeared near 460 n.m. and was followed by another band near 480 n.m. The appearance of these hypsochromic transitions was accompanied by formation of a bathochromic J-band with a principal absorption at 568 n.m. and a shoulder near 577 n.m. At relatively low mica/dye ratios (Curves a-c), the spectra pass through isosbestic points, an observation which suggests that the free dye is in equilibrium with dye adsorbed in its dimeric and its J-state. However, the isosbestic points are not maintained at the highest mica concentrations where the hypsochromic bands lose absorbance and definition and give way to a single intense J-band at 568 n.m. ($\epsilon = 2.3 \times 10^5 M^{-1}$ cm.$^{-1}$). This band has a half-width of about 13 n.m. and is associated with a weak subsidiary maximum near 520 n.m. The character of both transitions strongly resembles that obtained on natural, freshly-cleaved mica (63).

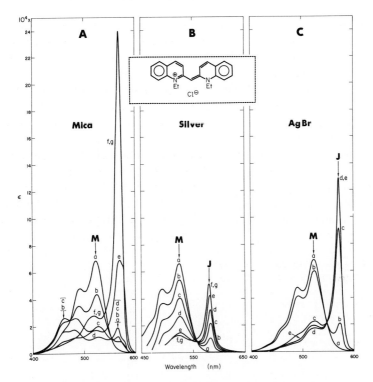

Figure 1. Transmission spectra at 23°C. of a fixed concentration of Pseudocyanine and varying amounts of aqueous dispersions. Dispersions without dye served as spectral references

A. 10^{-5}M dye in water: (a) 0.0 gram, (b) 0.008 gram, (c) 0.016 gram, (d) 0.064 gram, (e) 0.4 gram, (f) 2.0 gram, (g) 2.8 gram synthetic mica per liter

B. 4×10^{-5}M dye in 1% gelatin at pH 6, pBr 5.3: (a) 0.0 gram, (b) 0.1 gram, (c) 0.2 gram, (d) 0.3 gram, (e) 0.4 gram, (f) 0.6 gram, (g) 0.7 gram colloidal silver per liter

C. 5×10^{-6}M dye in 0.24% gelatin at pH 6.5, pBr 3: (a) 0.0 gram, (b) 0.02 gram, (c) 0.12 gram, (d) 0.22 gram, (e) 0.87 gram AgBr (Lippmann) per liter

In contradistinction to the spectra obtained in the mica dispersion, Pseudocyanine, when added to the colloidal silver and silver halide systems of Figures 1B and 1C, yielded no *H*-bands but formed well defined *J*-bands. The latter also exhibited a weak secondary peak located near the absorption maximum of dissolved, unperturbed dye. Although the position and intensity of the *J*-band varied with the substrate (*4, 17, 38, 61*), the differential spectra obtained in these silver systems exhibited marked similarities. An increase in the concentration of the substrate produced in both cases a monotonic change in the absorbance of the *M*-

and J-bands; whereas the former decreased, the J-bands increased in intensity until a maximum value was reached. Addition of further substrate had then no noticeable effect on the absorptivity of this band. These spectral changes, which were accompanied by the appearance of isosbestic regions near 555 n.m. with the silver, and near 545 n.m. with the silver bromide dispersion, suggested the existence of a substrate dependent equilibrium between two states: Unperturbed dye in solution and dye adsorbed in its J-state.

Quite parallel changes occur when a solution of the sulfate ester of poly(vinyl alcohol) (52) is added to dilute aqueous Pseudocyanine. As illustrated in Figure 2, there is no evidence for the formation of a dimer; one observes again the appearance of a J-band at 568 n.m. At the highest polymer concentration all unperturbed dye had apparently disappeared (Curve c) and further polymer addition did not change the shape of

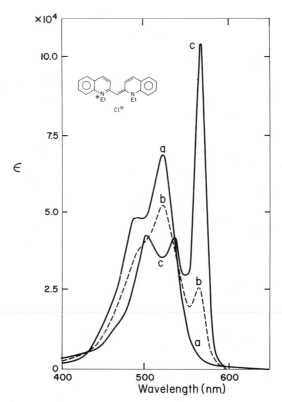

Figure 2. Transmission spectra at 23°C. of aqueous Pseudocyanine with varying concentrations of poly(vinyl alcohol) sulfate ester: (a) 0.0 gram, (b) 0.004 gram, (c) 0.004 gram PVA-sulfate per liter

this curve so that this spectrum represents the attainable conversion of the M- to the J-state. (It cannot be assumed that Curve c represents complete conversion of the dye to its J-state. Curve analysis by our colleague, F. Webster, has demonstrated that as much as 54% of the dye may have remained in its unperturbed M-state. Reconstruction of Curve c on that basis does not change its principal features but decreases the intensity of the maxima near 500 and 535 n.m.) This state manifests itself by the weak but resolved transitions near 500 and 535 n.m. and the intense band at 568 n.m. with an extinction greater than $1.2 \times 10^5 M^{-1}$ cm.$^{-1}$ and a half-width of about 15 n.m. Essentially identical results were reported for the spectra obtained with polyethylene sulfonate and anionic polysaccharides by Appel and Scheibe (1). In agreement with their deductions, present results indicate that salt formation between the cationic dye and closely adjacent anionic sites in the polymer is responsible for the spectra illustrated in Figure 2. Thus, no J-band was obtained with poly(vinyl alcohol) itself nor with a sulfonated polystyrene. In the latter polymer the distance between acid sites apparently was greater than the ca. 5 A., which appears to be the maximum distance permitting the J-state by electronic coupling between adjacent dye molecules (1). The importance of salt formation in this system is also illustrated by the fact that PVA-sulfate gave no J-band with a sulfoalkyl pseudocyanine which had a net charge of zero. It was noted, however, that in the silver and silver halide dispersions this zwitterionic dye exhibited adsorption and spectral properties that were essentially identical with those of Pseudocyanine itself.

The procedures used to obtain Figure 1 can be reversed. Figure 3 illustrates the spectral changes accompanying the addition of increasing amounts of Pseudocyanine to a fixed concentration of colloidal silver. The resulting family of differential spectra show that, after the J-band had reached a critical intensity (λ_{max} varied between 578-582 n.m.), additional dye exhibited only the spectral chracteristics of unperturbed dye in its solution state (cf., Figure 1B, Curve a). The salient features of these spectra are made more obvious by plotting the concentration of added dye against the absorbance of the corresponding J-band. This was done in Figure 4A. Following an initial linear increase in the absorbance of the J-band, a critical region was reached beyond which further addition of dye had little effect on this transition and caused a marked change in the slope of the curve. (The monomeric absorption spectra of cyanines often extend weakly but measurably into the J-region and even unperturbed Pseudocyanine in solution will make a slight contribution to J-band absorbance. Moreover, as already discussed, under some conditions of salt or surfactant concentrations, cyanines may exhibit intense J-bands in solution.) A change in slope at the same critical dye concentration was

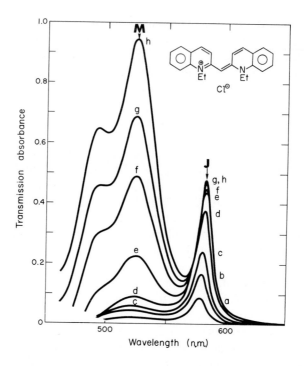

Figure 3. Transmission spectra of 4.65×10^{-3}M colloidal silver in 1% gelatin at 23°C., pBr 5.3, pH 6.5 with varying concentrations of Pseudocyanine: (a) 0.8, (b) 1.6, (c) 2.4, (d) 4.0, (e) 5.6, (f) 7.6, (g) 9.0, (h) 11.0×10^{-5}M dye. Undyed colloidal silver served as reference, 2mm. cells were used

also observed when the ratio of M to J absorbances was plotted against the concentration of added dye. It should be noted that absorbance in the M-band region involves not only unperturbed dye but also dye in its J-state which contains a component that absorbs at shorter wavelengths.

We considered the concentration-dependent spectral data of Figure 4A to be a manifestation of the adsorption equilibrium of the dye. Specifically, we assumed that the initial linear increase of J-absorbance represents binding of all the added dye at the substrate and that departure from this linear portion corresponds to added but unadsorbed dye; the latter would then account for the sudden increase of the M/J absorbance ratio. It follows from this interpretation that the intercepts obtained by extrapolating the linear sections of the plot of dye concentration *vs.* J. or M/J absorbance, should correspond to the amount of dye adsorbed at saturation of the surface. If we assume further that the absorbances of Figure 3 and 4A obey Beer's law, then the intensity of the J-band will be proportional to the surface concentration of the dye and the spectral data of Figure 4A can be converted into an adsorption isotherm. This was done in Figure 4B, where the amount of dye adsorbed per mole of colloidal silver (a) is plotted against the equilibrium concentration of dye in solution (c). The resulting isotherm is expressed by the Langmuir adsorption law

$$\frac{c}{a} = \frac{c}{a_s} + \frac{1}{a_s K}$$

where a_s is the saturation coverage and K is the Langmuir adsorption coefficient. The slope of the linear relation of c/a against c yielded a

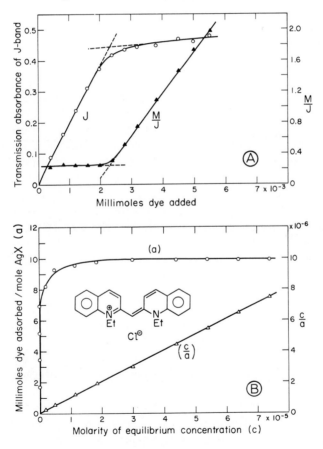

Figure 4A. *Absorbance of the J-band and the M/J ratio at varying concentrations of Pseudocyanine in 50 ml. of 4.65 mM colloidal silver (cf. Figure 3)*

B. *Adsorption isotherm of Pseudocyanine in colloidal silver calculated from the spectral data of Figure 4A. See text for details*

saturation coverage, a_s, of 10 millimoles of Pseudocyanine per mole of silver. On the assumption that the packing of the dye at this silver surface is similar to that on silver halide substrates—*i.e.*, 57A.² per molecule (*23*), a specific area of 35 sq. meter per gram of silver was obtained. A similar analysis of the spectral data of Figure 1B gave the same results. If the colloidal silver particles are considered to be spherical, this surface area is equivalent to an average particle diameter of 180 A. This value agrees in magnitude with diameter estimates of 110, 140, and 190 A. obtained by Stevens and Block (*64*) with three independent methods on similarly prepared silver dispersions. Moreover, application of the described spectral method to a different silver dispersion yielded an average particle diameter of 45 A., whereas a value of about 48 A. was obtained from electronmicrographs. This agreement with independent size estimates makes it probable that the assumptions applied to the interpretation of the dye spectra in silver dispersions were valid.

As already indicated, the spectra of Pseudocyanine in the mica dispersion (Figure 1A) would require a more detailed analysis for their quantitative interpretation. In common with silicates like montmorillonite (*3*), the synthetic mica behaves as if it possesses different types of binding sites which caused adsorption of the dye both in the dimeric and in the *J*-state. In contrast to the mica-dye system, transmission spectra of Pseudocyanine in the colloidal silver bromide indicated the predominance of only one type of surface interaction (Figure 1C). Analysis of these spectra by the method described for the case of colloidal silver, gave a specific area of 11 sq. meter/gram AgBr with an average particle diameter of 950 ± 60 A. The accuracy of these results is uncertain since the true particle dimensions were not known (*cf., Experimental* section).

The applicability of this *in situ* method for the determination of surface areas depends not only on knowledge of the dye's molecular area in the adsorbed state but also on the assumption that the chosen spectral parameter measures the surface concentration of the dye. In order to test the relation between adsorption of dye to silver halide and its spectral characteristics in the bound state, the behavior of Pseudocyanine in a coarse silver halide suspension (Dispersion D) was studied. This particular dispersion was chosen because some of its relevant adsorption characteristics had already been examined (*22, 23*). Moreover, observations by Boyer and Cappelaere with Pseudocyanine adsorbed on AgBr powders (*5*) indicated that *J*-band intensity varied with the amount of adsorbed dye and was not sensitive to the concentration of Ag^+ or Br^- ions in the range pAg 3.3-8.7.

There remained the question of how to evaluate dye spectra in concentrated silver halide dispersions whose particle sizes ranged between 0.2-2.0 μ, and where multiple interparticle light scatter made it unfeasible to apply transmittance measurements of the kind used in Figures 1 and 3. In other studies (*46, 50, 70, 71*) the following relation,

% Absorbance = 100 − % Transmittance − % Reflectance

or one of its components, had been used. However, in the absence of calibration data, the results did not yield a verifiable direct relation between the photometric parameters and the surface concentration of dye. In this work we applied the Kubelka-Munk reflectivity function, where K and S are absorption and scatter coefficients of the substrate, R_∞ is the reflectivity of an infinitely thick layer, and C and ϵ are the molar concentration

$$K/S = \frac{2 R_\infty}{(1 - R_\infty)^2} = C \frac{\epsilon}{S}$$

and molar absorptivity coefficients, respectively. This relation, which has been studied in much detail by Kortüm, who also applied it to adsorption phenomena (*31, 32, 33*), has a form similar to Beer's law, except that in reflectometry the concentration C is proportional to K/S rather than to absorbance.

The Kubelka-Munk function is applicable to silver halide dispersions, provided that specular or Fresnel reflections are negligible and that infinite reflectivities, R_∞, are actually measured (*24*). After it had been ascertained that these conditions had been met, a concentration series of Pseudocyanine in the coarse silver bromide suspension (Dispersion D) was prepared and yielded the K/S spectra illustrated in Figure 5A. It is apparent that these concentration-dependent reflection spectra possess characteristics similar to the transmission spectra obtained with the silver substrate in Figure 3. Again one observes that the *J*-band near 572 n.m. tends toward a maximum value. In Figure 5B the millimoles of dye added per mole of silver bromide are plotted against *J*-band intensity expressed in K/S values. The character of the resulting curve is similar to the corresponding graph in Figure 4A and the data were treated similarly. Extrapolation of the linear parts in Figure 5B yielded an intercept which was again taken to define the amount of dye adsorbed at saturation coverage and served as the reference point for the conversion of the spectral data into the adsorption isotherm in Figure 8A.

In comparison with an azo dye adsorbed on anhydrous barium sulfate (*31*), the K/S values of Figure 5 are remarkably large. In part this is because of the high extinction coefficient of the Pseudocyanine's *J*-band (*cf.*, Figure 1) and in part to the non-zero base line in Figure 5A. It was experimentally convenient to use an 0.2 absorbance filter in the sample beam of the spectrophotometer; in the absence of this filter—it had no influence on the final results—the K/S value at saturation coverage would

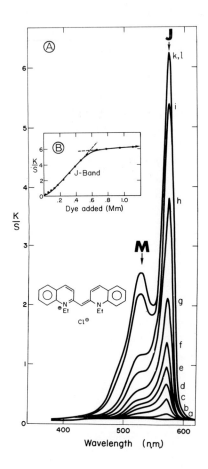

Figure 5A. Reflection spectra of 6.43×10^{-2}M AgBr (Dispersion D) in 0.2% gelatin at 23°C., pBr 3, pH 6.5 with varying concentrations of Pseudocyanine chloride (a) 0.01, (b) 0.05, (c) 0.25, (d) 0.5, (e) 0.75, (f) 1.0, (g) 1.5, (h) 2.5, (i) 3.5, (k) 6.0, (l) 7.0×10^{-5}M dye. Sample beam of spectrophotometer passed through a 0.2 absorbance filter, undyed AgBr served as spectral reference, 40 mm. cells were used. See text for explanation of reflection parameter K/S

B. Dependence of J-band reflectivity on concentration of Pseudocyanine. Millimoles of dye in 100 ml. of 6.43×10^{-2}M AgBr

be 3.2 instead of near 6.0. In Figure 5B a slight toe is noticeable in the curve at low dye concentrations. To the extent that this is caused by a systematic reflectivity error it may be corrected (*32, 34, 37*), but for present purposes there was no need to consider this departure from linearity at low surface coverage by the dye.

The very same dispersions which had yielded the spectra of Figure 5A were centrifuged to separate solid and liquid phases in order to measure dye adsorption by conventional methods (*23, 70*). The resulting data are also shown in Figure 8A where they can be compared with the spectrally determined adsorption isotherm. The agreement between the two independent methods is good and indicates that *in situ* spectra can be used to determine directly the surface concentration of Pseudocyanine. This dye is particularly suitable for that purpose because its spectra at elevated surface coverages always seem to be associated with the narrow J-band which has a high extinction coefficient.

However, the formation of a J-band is not a prerequisite for the spectral determination of surface concentrations of a dye; it is sufficient that the difference between solution and surface spectra can be evaluated quantitatively. Astraphloxin is an example of a dye that meets these conditions. In dilute aqueous solution its principal absorption at 542 n.m. is separated some 1250 cm.$^{-1}$ from a subsidiary and probably vibrational maximum at 521 n.m. At concentrations in excess of $2 \times 10^{-4}M$, this dye exhibits a new hyposochromic transition at 505 n.m. considered to be caused by a dimer (*12*); no J-band has been found in Astraphloxin solutions under conditions where Pseudocyanine does form one. Figure 6 illustrates the spectra obtained by a concentration series of Astraphloxin in the same silver bromide suspension (Dispersion D) and under the same conditions as had been used in connection with Figure 5. However, instead of plotting the data as K/S values, we have recorded them as the more accessible reflection absorbances ($-\mathrm{Log}\, R_\infty$) from which % R_∞ and K/S could be computed for any desired wavelength. In order to facilitate comparisons, Figure 6 also contains a transmittance spectrum of Astraphloxin (dashed curve) in the absence of the silver halide. In accord with the difference in refractive index of water and AgBr (*72*), the spectrum of the dye at low surface coverages is displaced approximately 25 n.m. towards longer wavelengths (Curve a, b). Hence, the new principal transition near 565 n.m. is designated the M_a-band because it is associated with isolated, unperturbed but adsorbed molecules. As the dye concentration is increased, two other transitions become defined in Figure 6, a hypsochromic peak (*H*) near 530 n.m. and a bathochromic band (*B*) near 580 n.m. At the most elevated dye levels, the B-band reaches a limiting value whereas the H-band shifts towards the maximum associated with the free dye in solution (542 n.m.). At first it was thought

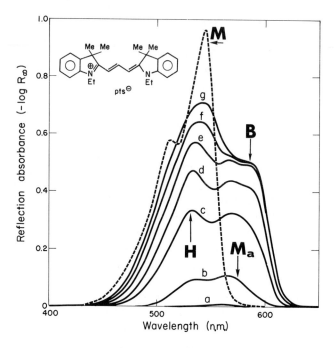

Figure 6. Reflection spectra of 6.26 × 10⁻²M AgBr (Dispersion D) in 0.2% gelatin at 23°C., pBr 3, pH 6.5 with varying concentration of Astraphloxin: (a) 0.01, (b) 0.10, (c) 1.0, (d) 2.0, (e) 3.0, (f) 4.0, (g) 5.0 × 10⁻⁵M dye. Undyed AgBr served as spectral reference, 40 mm. cells were used. Dashed curve: Transmission absorbance of 7 × 10⁻⁶M dye (1 cm. cell) in same solvent system after removal of AgBr by centrifugation

that the H- and B-bands might represent two different states of adsorbed Astraphloxin. However, experiments analogous to those in Figure 1, where dye concentration was kept constant but AgBr was added in various amounts, yielded spectra which suggested a different interpretation. Over a limited range of AgBr, a family of spectra resulted which, in addition to the characteristic transitions associated with free dye, exhibited H-, and B-bands. All these spectra passed through an isosbestic point (558 n.m.); thus they not only indicated an equilibrium between free and adsorbed dye but also suggested that the cited bands are a manifestation of a single state of the adsorbed dye. (When Astraphloxin is adsorbed on 100-faces of cubic AgBr rather than on its octahedral 111-faces, the transitions of the B-state are displaced bathochromically by about 5 n.m. Moreover, in some silver halide systems, shoulders near 490 and 610 n.m. can be noted; the latter has been described as a J-band (54).) Following West et al. (46), this condition will be referred to as the B-state and is assumed to be a consequence of perturbation between neighboring dye molecules adsorbed and packed closely on their long axis. Unlike the similarly arrayed J-state, it is thought that the principal transition of the B-state involves an electronic moment parallel to the conjugated chain of the molecule. Either the B- or the M_a-band can be used to calculate the surface saturation by Astraphloxin. Despite overlapping absorption from free dye, the M_a-band has the advantage that it measures contributions from adsorbed, isolated molecules at low coverage as well as contributions from the perturbed B-state which is observed at higher surface concentrations. Hence, by using differential reflection spectra, some of which are shown in Figure 6, the concentration-dependence of the M_a-band was obtained and is expressed in Figure 7 in terms of various photometric parameters. It can be seen that neither % R_∞ nor $-\text{Log } R_\infty$ (reflection absorbance) produced a direct relation from which the dye's surface concentration could be estimated. On the other hand, the K/S reflectivity parameter again yielded two linear plots whose extrapolated intercept furnished the amount of dye adsorbed at saturation coverage. As before, this made it possible to convert the K/S data into an adsorption isotherm which was again found to be in accord with independently obtained adsorption determinations (Figure 8A).

The adsorption isotherms of Astraphloxin and Pseudocyanine in Figure 8A were also expressed by the Langmuir adsorption equation. The resulting linear relation of Figure 8B demonstrates excellent agreement between the data obtained from phase separation and spectral techniques.

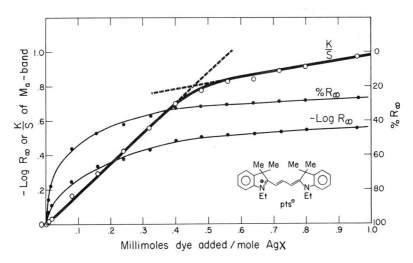

Figure 7. Dependence of M_a-band reflectivity of Astraphloxin in Figure 6 expressed in terms of different reflection parameters. See text for significance of K/S values

From the slope and intercept of these plots the saturation coverage and adsorption coefficient were obtained by application of the previously given equation. The results are listed in Table I, which also includes the area per dye molecule at saturation coverage and the standard free energy of adsorption, $\Delta G°$. Although the latter parameter was calculated as before (23), its thermodynamic validity is questionable since reversibility of adsorption of these dyes was not demonstrated. The molecular

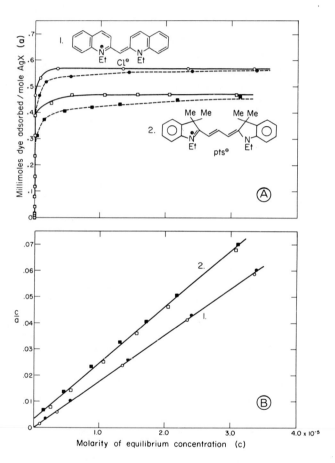

Figure 8A. Adsorption isotherms of Pseudocyanine (No. 1, Circles) and of Astraphloxin (No. 2, Squares) in AgBr (Dispersion D) containing 0.2% gelatin at 23°C., pBr 3, pH 6.5. The data are expressed as the concentration of free dye (c) in equilibrium with dye adsorbed per mole of AgBr (a). Open data points and solid lines: Results calculated from surface spectra. Solid data points and dashed lines: Results obtained from phase-separation procedure

B. Adsorption isotherms of Figure 8A expressed in terms of the Langmuir equation. See text

area of Astraphloxin was obtained from its saturation coverage relative to that of Pseudocyanine to which an area of 57 A.^{2} had been assigned. As discussed elsewhere (23), a value of this magnitude is not only consistent with the theoretical dimensions of the dye bound on its long edge in a closely packed array, but it is also in reasonable agreement with various cited measurements obtained in other laboratories.

Adsorption determinations with Astraphloxin in homodisperse cubic and octahedral silver bromide dispersions of the type which were previously described (20, 23), indicated a mean limiting area for this dye of 76 ± 7 A.^{2} which is in fair agreement with the 69 A.^{2} obtained from Figure 8. Since a monolayer of dye in its flat orientation would require an area in excess of 130 A.^{2} per molecule, the results indicate that at saturation Astraphloxin was adsorbed at its long edge (ca. 19 A.) with an intermolecular spacing of about 4.0 A. As seen in the spectra, adsorption occurred without formation of a J-state and its absence in these experiments may account for its relatively low free energy of adsorption (Table I). It must be stressed that, although the *gem*-dimethyl groups of Astraphloxin are located perpendicular to the conjugated chain, close spacing of molecules is still possible if they are oriented on their long edge in a staggered array. Presumably, it is this orientation which, contrary to Sheppard's early supposition (61), is responsible for the formation of J-bands in the N,N'-dimethyl analog of this dye (54) and in its derivatives having substituents in the benzene ring.

The study of surface-dye interaction by the *in situ* optical procedure has also been applied to 5,5'-dichloro-3,3',9-triethylthiacarbocyanine, which was known to be adsorbed in multilayers (70). When this dye was added to a suspension of octahedral AgBr at pBr 3, a J-band was formed near 640 n.m. (Figure 9A). Increasing dye concentrations enhanced the reflectivity of this band until it reached a limiting intensity; once reached, additional dye contributed insignificantly to reflectivity values in that spectral region. (However, overlapping absorption from dye in other states may, at high dye concentrations, cause an apparent contribution to absorption. This effect is illustrated by the highest dye level shown in Figure 9A.) As before, the concentration-dependent reflection densities of the J-band were converted to K/S functions (Figure 9B) from which the adsorption isotherm marked "Optical Method" was then derived (Figure 9C). The same figure also shows an adsorption isotherm obtained from the same system by the classical phase separation technique. The latter measurement exhibits the slight horizontal step which West *et al.* had interpreted as signifying completion of a monolayer before the onset of polylayer adsorption (70).

Striking differences are apparent between the two independently determined adsorption isotherms which are illustrated in Figure 9C.

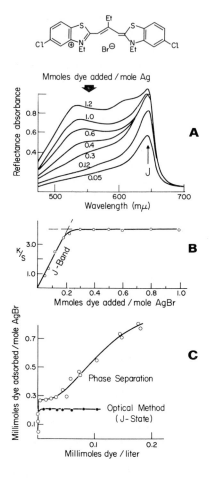

Figure 9A. Reflection spectra of 0.1M octahedral AgBr (Dispersion C of Reference 23) in 0.3% gelatin at 23°C., pBr 3, pH 6.5 with added 5,5'-dichloro-3,3'-9-triethyl-thiacarbocyanine bromide in millimoles per mole AgBr

B. Dependence of J-band reflectivity at 640 n.m. on concentration of the thiacarbocyanine in the AgBr dispersion. See text for significance of K/S values

C. Adsorption isotherms of the thiacarbocyanine in the octahedral AgBr dispersion. Open data points: Results calculated from surface spectra of Parts A and B. Solid data points: Results obtained from phase-separation procedure

Whereas the optical method indicates attainment of saturation coverage by the dye in its J-state, results obtained with the phase-separation technique show that, after reaching apparent saturation, further adsorption occurs as the dye concentration in solution is increased. The horizontal step observed with the phase separation measurement is in approximate agreement with the maximum surface concentration of dye in its J-state as determined by the optical method. Hence, it is concluded that, contrary to earlier suppositions (70), only the first layer of this dye is adsorbed in its J-state; subsequent dye layers must be adsorbed in different states.

Table I. Adsorption Data for Cyanines on Silver Bromide
7×10^{-2}M AgBr (Dispersion D) in 0.2% gelatin at 23°C., $pH = 6.5$, $pBr = 3$

	Pseudocyanine	Astraphloxin
Saturation Coverage	0.57[a]	0.46[a]
(mM Dye/mole AgX)	0.56[b]	0.47[b]
Area per molecule (A.^{2})	57	69
Langmuir Coefficient K (mM^{-1})	3×10^3	7×10^2
$\Delta G°$ (kcal./mole)	−9.8	−8.8

[a] Values obtained from phase-separation procedure.
[b] Values calculated from spectra of Figures 5 and 6.

Discussion

On comparing the Pseudocyanine spectra of Figures 1, 2, 3, and 5 with those obtained in concentrated aqueous dye solutions (12, 56, 57, 58, 59), obvious similarities of the resulting J-states will be noted. Present and earlier data make it apparent that electronic coupling between adjacent dye molecules can produce similar J-states, regardless of whether appropriate orientation and proximity of the molecules is caused by Coulombic attraction, as in salt formation with polymeric ions, or whether it is induced by van der Waals forces. The latter are primarily involved in the formation of dye aggregates in water as well as in the charge-independent adsorption of dye monolayers at silver halide surfaces (23, 46).

The principal spectral feature of the J-state of Pseudocyanine is the bathochromic band which, depending on the substrate, absorbs between 568-582 n.m. with an extinction coefficient of 0.5-2.3 $\times 10^5 M^{-1}$ cm.$^{-1}$ and a half-width of less than 50 cm.$^{-1}$. These values are in rough agreement with those obtained in aqueous gels of Pseudocyanine at concentrations above 0.1M (12). In addition to this intense and narrow band, the J-state also contains components which absorb weakly at shorter wavelengths. This spectral splitting of the J-state has been discussed with reference to coupled oscillators and to quantum theory by Förster; more recently, McRae and Kasha have treated this subject in terms of the molecular exciton model (*16, 45*). The low-intensity transitions of the J-state overlap the absorption of unperturbed dye in solution; they are poorly defined in Figure 1, where they appear as shoulders extending from about 480-540 n.m. However, they are clearly resolved in the spectra resulting from interaction with appropriate anionic polymers and exhibit distinct subsidiary maxima of 500 and 535 n.m. (Figure 2, and Reference 1). These short-wave components can also be observed on examining the anisotropic behavior of fibrous J-aggregates in water, where they appear near 500 and 540 n.m. with the electric vector perpendicular to the fiber axis and parallel to the longest dimension of the molecule (*25, 41, 56, 57, 58, 59*). Such high-frequency bands were found to be associated as an integral part of the J-state with all cyanines which were examined.

Apparently, it was largely lack of recognition of these short-wave components of the J-state which recently led Padday and Wickham to the conclusion that in silver halide dispersions Pseudocyanine is adsorbed at saturation coverage in two dye layers. It was supposed that absorption in the M-band region (540 n.m.) was caused only by a flat monolayer of dye covering the surface and that a second dye layer was bound on top of the first in an edge-on orientation which then gave rise to the J-band at 573 n.m. (*50*). These views concerning an *a priori* unlikely dye configuration have received no experimental support from earlier or present evidence. However, in agreement with conclusions summarized by Carroll (*8*), present experiments have shown that adsorption of isolated Pseudocyanine could be detected by the appearance of an M_a-band only under conditions of very low surface coverage. The M_a-band at 543 n.m. in Figure 5 (Curve a) is barely evident before the onset of distinct J-band formation at higher dye concentrations which presumably involves a reorientation of the dye molecules (*10, 70*). The extent to which M_a-bands form appears to depend also on the crystal habit of the silver halide. Thus, continuing the previously reported adsorption measurements of Pseudocyanine in crystallographically defined AgBr gelatin dispersions (*23*), it was found that, in the cubic AgBr system, the dye was adsorbed in its M_a-state at coverages approaching 5% of the surface before the J-state became prevalent. On the other hand, in the octahedral AgBr dispersion no M_a-band could be detected and only the J-band was observed. This dependence of M_a-state concentration on the crystal habit of the substrate is believed to indicate differences in their adsorption forces. It is likely that, where substrate-dye interaction is strongest, the dye will be adsorbed as isolated and possibly flat molecules (*10, 70*), whereas lateral dye-dye interaction leading to the *J*-state will be favored when competing forces from the substrate are relatively weak. According to this interpretation, the energy of adsorption of Pseudocyanine at low coverages of AgBr should be greater for the cubic than for the octahedral substrate.

The silver dispersion gave no indication that Pseudocyanine was bound in the M_a-state at any surface coverage (Figures 1B and 3) and in this respect it behaved like the octahedral AgBr substrate. Although colloidal silver having reproducible spectral properties was readily prepared by the indicated procedure, these silver preparations are neither homodisperse, nor do they possess a well-defined surface. A more suitable silver substrate was prepared by depositing silver on gold nuclei (*48*). This sol was purified by dialysis; it had a narrow size distribution with an average particle diameter of 320 A. and on centrifuging gave an optically clear supernatant liquid. Under the same conditions as those described in Figure 1B, Pseudocyanine again gave a *J*-band near 580 n.m.; however, its molar extinction was twice as high as that shown in Figure 1B. More detailed examination of the interaction of this silver sol with the *p*-toluenesulfonate salt of Pseudocyanine by both the centrifugation and spectral procedure, demonstrated that adsorption of the dye and *J*-band formation occurred in the presence but not in the absence of chloride, bromide, or iodide. Furthermore, if reducing compounds like hydrazine or silver-complexing agents such as sulfite or cyanide were added to the system before the dye, neither *J*-band formation nor adsorption was detected, even in the presence of $10^{-4}M$ or more halide. Hence, we tentatively conclude that the adsorption sites for Pseudocyanine in these dispersions are not silver but consist of silver halide. In this connection it is significant that previously reported adsorption of this dye on silver substrates was observed on addition of halide salts of the dye (*19, 49*).

Present data illustrate the technique for an *in situ* determination of surface areas. Related methods had been applied primarily to the study of site distributions in clay minerals, particularly by Russian workers (*66*), and they were used by Bergmann and O'Konski in a detailed investigation of the methylene blue-montmorillonite system (*3*). In fact, changes in electronic spectra arising from surface interactions received sufficient attention in the past to warrant their review by A. Terenin (*65*). Most of these investigations involved transmittance spectra but new techniques in reflection spectrophotometry and applications of the Kubelka-Munk relation have facilitated the quantitative evaluation of spectra in highly turbid media (*35, 69, 77*). Thus, in agreement with the work of Kortüm on powders and anhydrous dispersions (*31, 32, 33*), our results demonstrate the applicability of the Kubelka-Munk function to strongly scattering aqueous silver halides. They also show that the perturbation of a dye's electronic transitions induced by its adsorption can yield direct information on surface coverages (24) and allows conclusions to be made about molecular orientation in the dye monolayer. For the purpose of measuring the surface area of a substrate by such spectral changes, the use of Pseudocyanine commends itself. Not only is the molecular area of this dye reasonably well known but the narrow and intense transition associated with its close-packed adsorbed state does not depend strongly on the composition of the silver halide or its crystal habit. Moreover, present results and earlier experiments have demonstrated that dye adsorption measurements with Pseudocyanine, either in the absence or presence of protective colloids, can yield silver halide surface areas which are in good agreement with various independent estimates (9, 23).

It remains to be determined to what extent the dye adsorption technique is applicable to other substrates. No evidence was obtained for Pseudocyanine adsorption to MnO_2, Fe_2O_3 or to pure silver surfaces, although this dye can be bound to mica, lead halides, and mercury salts with formation of a J-band (61). Not only cyanines but other dye classes can yield surface spectra which may be similarly analyzed. This is specifically the case with the phthalein and azine dyes which were recommended by Fajans and by Kolthoff as adsorption indicators in potentiometric titrations (15, 30). The techniques described are also convenient for determining rates and heats of adsorption and surface concentrations of dyes; they have already found application in studies of luminescence (18) and electrophoresis (68) of silver halides as a function of dye coverage.

Sorption of Meso-alkylcarbocyanines

Although with Pseudocyanine the existence of a J-state does not depend on the silver halide's crystal structure, meso-alkylcarbocyanines can exhibit remarkably different surface spectra on cubic and octahedral silver bromides (13, 17). These dyes are particularly sensitive to variation in spacing of lattice sites at (100) and (111) faces and to the different field forces emanating from them. Thus, with variously charged 9-methylthiacarbocyanines (22, 23, 36), further experiments at pBr 3-4 in cubic AgBr dispersions showed H- and B-bands of essentially equal intensity near 520 and 590 n.m., respectively. These bands were nearly absent on the octahedral substrate which, instead, caused formation of a J-band near 620 n.m. Of the various possibilities to account for these spectral differences, there are at least three which warrant close consideration: (a) The thiacarbocyanine may be adsorbed either flat or on its long edge, depending on the substrate. (b) Stacking of dye molecules on their long edge may vary in orientation such that the nitrogen atoms of the dye can be near the surface in one case, whereas in the other, sulfur would be in that position. Alternating orientations of this type occurred with a thiacarbocyanine (75, 76) in its crystalline state. (c) Different stereoisomers of the dye are adsorbed on the two substrates. The existence of interconvertible stereoisomers of 9-alkylthiacarbocyanines has been demonstrated (60, 74) and Zechmeister has reported the chromatographic separation of a thiatricarbocyanine into such isomers (78).

Of these alternatives only the first can now be eliminated since the cationic 9-methylthiacorbocyanine was found to occupy the same limiting area of ca. 60 A.^{2} per molecule in both cubic and octahedral AgBr (23). For a close-packed monolayer, this area requirement is consistent with essentially vertical alignment but is too small, by a factor greater than two, to allow for horizontal orientation of the adsorbed dye. However, relatively minor variations in tilt angle of the molecular planes, e.g., 50° vs. 60° would probably not be detectable by these area determinations. Yet such angular displacements (14, 45) are expected to exert pronounced effects on surface spectra. Hence, it is clear that sufficient data are not yet available to fully support an explanation of spectral properties of meso-substituted carbocyanines which are bound to silver halides having different crystal habits.

Acknowledgments

It is a pleasure to thank H. Elins, for having brought to our attention the interaction of Pseudocyanine with PVE-sulfate; F. Grum, for assistance in the determination of spectra; and J. Helling, for much help in early adsorption measurements.

Note

The area determinations by dye adsorption from solution discussed here are applicable to aqueous dispersions. Although saturation coverage of silver halides by Pseudocyanine remained unchanged in 40% methanol by volume, it is known that in organic solvents where ion-pairs may be adsorbed, the molecular cross section of the cyanine can vary with the dye's anion—cf. Reference 23 for discussion and literature citations. Recent determinations of AgI areas by adsorption of Pseudocyanine were reported to have been unrealistic and salt-dependent (van den Hul, H. J., Lyklema, J., *J. Phys. Chem.* **90**, 3010 (1968)). A likely reason for this result is the circumstance that these investigators carried out their measurements in alcohol dispersions of the substrate where the cited solvent-dependent limitations would apply.

Literature Cited

(1) Appel, W., Scheibe, G., *Z. Naturforschg.* **13b,** 359 (1958).
(2) Bean, R. C., Shepherd, W. C., Kay, R. E., Walwick, E. R., *J. Phys. Chem.* **69,** 4368 (1965).
(3) Bergmann, K., O'Konski, C., *J. Phys. Chem.* **67,** 2169 (1963).
(4) Borginon, H., Danckaert, V., *Phot. Korr.* **98,** 74 (1962).
(5) Boyer, S., Cappelaere, J., *J. Chim. Phys.* **60,** 1123 (1963).
(6) Boyer, S., Pinchon, L., Degove, O., *J. Chim. Phys.* **60,** 301 (1965).
(7) Brooker, L. G. S., Keyes, G. H., *J. Am. Chem. Soc.* **57,** 2488 (1935).
(8) Carroll, B. H., *Phot. Sci. Eng.* **5,** 65 (1961).
(9) Curme, H. G., Natale, C. C., *J. Phys. Chem.* **68,** 3009 (1964).
(10) Davey, E. P., *Trans. Faraday Soc.* **36,** 323 (1940).
(11) Dickinson, H. O., *Nature* **163,** 485 (1949).
(12) Ecker, H., *Koll. Z.* **92,** 35 (1940).
(13) Eggers, J., Gunther, E., Moisar, E., *Phot. Korr.* **102,** 144 (1966).
(14) Emerson, E., Conlin, M., Rosenoff, A., Norland, K., Rodriguez, H., Chin, D., Bird, G., *J. Phys. Chem.* **71,** 2396 (1967).
(15) Fajans, K., "Newer Methods of Volumetric Chemical Analysis," trans. by R. Oesper, D. Van Nostrand Co., New York, 1938.
(16) Förster, Th., "Fluoreszenz Organischer Verbindungen," p. 256 ff, Vandenhoeck and Ruprecht, Goettingen, 1951.
(17) Frieser, H., Graf, A., Eschrich, D., *Z. Electrochem.* **65,** 870 (1961).
(18) Gilman, P. C., *Phot. Sci. Eng.* **11,** 222 (1967).
(19) Groszek, A., Wood, H., *J. Phot. Sci.* **13,** 133 (1965).
(20) Gunther, E., Moisar, E., *J. Phot. Sci.* **13,** 280 (1965).
(21) Hamer, F. M., *J. Chem. Soc.* **1928,** 206.
(22) Herz, A. H., Helling, J. O., *J. Colloid Sci.* **17,** 293 (1962).
(23) Herz, A. H., Helling, J. O., *J. Colloid Interface Sci.* **22,** 391 (1966).
(24) Herz, A. H., Helling, J. O., *Koll. Z. and Z. Polymere* **218,** 157 (1967).
(25) Hoppe, W., *Koll. Z.* **109,** 27 (1944).
(26) Jelley, E. E., *Nature* **138,** 1009 (1936).
(27) *Ibid.,* **139,** 631 (1937).
(28) Kay, R. E., Walwick, E. R., Gifford, C. K., *J. Phys. Chem.* **68,** 1896 (1964).
(29) *Ibid.,* **68,** 1907 (1964).
(30) Kolthoff, I. M., *Chem. Rev.* **16,** 87 (1935).
(31) Kortüm, G., Vogel, J., *Chem. Ber.* **93,** 706 (1960).
(32) Kortüm, G., Oelkrug, D., *Z. Physik. Chem. (Frankfurt)* **34,** 58 (1962).
(33) Kortüm, G., Braun, W., *Z. Physik. Chem. (Frankfurt)* **48,** 282 (1966).
(34) Kortüm, G., Oelkrug, D., *Naturwiss.* **53,** 600 (1966).
(35) Kottler, F., "Progress in Optics," Vol. III, p. 1, E. Wolf, ed., North Holland Publishing Co., Amsterdam, 1964.
(36) Kragh, A., Peacock, R., Reddy, G., *J. Phot. Sci.* **14,** 185 (1966).
(37) Lieu, V., Frodyma, M., *Talanta* **13,** 1319 (1966).
(38) Markocki, W., *J. Phot. Sci.* **13,** 85 (1965).
(39) Mason, S. F., *Proc. Chem. Soc.* **1964,** 119.
(40) Mason, S. F., "Optische Anregung Organischer Systeme," p. 143, 2nd. Intern. Farbensymposium, Verlag Chemie, Weinheim, Germany, 1966.
(41) Mattoon, R. W., *J. Chem. Phys.* **12,** 263 (1944).
(42) McKay, R. B., Hillson, P. J., *Trans. Faraday Soc.* **61,** 1800 (1965).
(43) *Ibid.,* **62,** 1439 (1966).
(44) *Ibid.,* **63,** 777 (1967).
(45) McRae, E., Kasha, M., "Physical Processes in Radiation Biology," p. 23 ff., Academic Press, New York, 1964.
(46) Mees, C. E. K., James, T. H., "The Theory of the Photographic Process," 3rd. ed., Chapt. 12, The Macmillan Co., New York, 1966.

(47) *Ibid.*, Chapt. 1.
(48) Morriss, R., Collins, L., *J. Chem. Phys.* **41**, 3357 (1964).
(49) Newmiller, R. J., Pontius, R. B., *Phot. Sci. Eng.* **5**, 283 (1961).
(50) Padday, J. F., Wickham, R., *Trans. Faraday Soc.* **62**, 1283 (1966).
(51) Padday, J. F., *J. Phys. Chem.* **71**, 3488 (1967).
(52) Patat, F., Vogler, K., *Helv. Chim. Acta* **35**, 128 (1952).
(53) Paticolas, W., *J. Chem. Phys.* **40**, 1463 (1964).
(54) Riester, O., "Scientific Photography," p. 480, Proc. Int'l. Colloq. Liege, 1959, H. Sauvenier, ed., Pergamon Press, London, 1962.
(55) Ruggli, P., Jensen, P., *Helv. Chim. Acta* **18**, 624 (1935).
(56) Scheibe, G., *Z. Angew. Chem.* **50**, 51 (1937).
(57) Scheibe, G., *Koll. Z.* **82**, 1 (1938).
(58) Scheibe, G., *Z. Elektrochem.* **52**, 283 (1948).
(59) Scheibe, G., "Optische Anregung Organischer Systeme," p. 109, 2nd Intern. Farbensymposium, Verlag Chemie, Weinheim, Germany, 1966.
(60) Scheibe, G., Worz, O., *Angew. Chem.* **78**, 304 (1966).
(61) Sheppard, S. E., *Rev. Mod. Phys.* **14**, 303 (1942).
(62) Sheppard, S. E., Lambert, R. H., Walker, R. D., *J. Chem. Phys.* **7**, 265 (1939).
(63) Skerlak, T., *Koll. Z.* **95**, 265 (1941).
(64) Stevens, G. W. W., Block, P., *J. Phot. Sci.* **12**, 247 (1964).
(65) Terenin, A., *Advan. Catalysis* **15**, 227 (1964).
(66) Vedeneeva, N., quoted by Eitel, W., "Silicate Science," p. 448, Vol. 1, Academic Press, New York, 1964.
(67) Weiser, H. B., "Inorganic Colloid Chemistry," Vol. 1, p. 119, J. Wiley and Sons, New York, 1933.
(68) Weiss, G. R., Ericson, R. H., Herz, A. H., *J. Colloid Interface Sci.* **23**, 277 (1967).
(69) Wendlandt, W., Hecht, H., "Reflectance Spectroscopy," Interscience Publishers, New York, 1966.
(70) West, W., Carroll, B. H., Whitcomb, D. L., *J. Phys. Chem.* **56**, 1054 (1952).
(71) West, W., Carroll, B. H., Whitcomb, D. L., *Ann. N. Y. Acad. Sci.* **58**, 893 (1954).
(72) West, W., Geddes, A. L., *J. Phys. Chem.* **68**, 837 (1964).
(73) West, W., Pearce, S., *J. Phys. Chem.* **69**, 1894 (1965).
(74) West, W., Pearce, S., Grum, F., *J. Phys. Chem.* **71**, 1316 (1967).
(75) Wheatley, P., *J. Chem. Soc.* **1959**, 3245.
(76) *Ibid.*, **1959**, 4096.
(77) Wolfe, R. N., DePalma, J. J., Saunders, S., *J. Opt. Soc. Am.* **55**, 956 (1965).
(78) Zechmeister, L., Pinckard, J., *Experentia* **9**, 16 (1953).

RECEIVED October 26, 1967.

15

The Relative Adsorbability of Counterions at the Charged Interface

KŌZŌ SHINODA and MASAMICHI FUJIHIRA

Yokohama National University, Ookamachi, Minamiku, Yokohama, Japan

The distribution of counterions at the charged surface and in the bulk solution was studied combining the radiotracer technique with foam fractionation technique. The relative adsorbability of counterions was determined from the ratio of mole ratios in an adsorbed and in the solution phases. The relative adsorbability was $Cl^- : CH_3COO^- : Br^- : ClO_3^-$; $NO_3^- = 1 : 0.7 : 1.5 : 3.0 : 4.1$ in solution of dodecylammonium(0.011 mole/liter), $Cl^- : Br^- : ClO_3^- : NO_3^- : SO_4^{2-} = 1 : 1.2 : 2.8 : 1.6 : 2.3$ in solution of dodecyltrimethylammonium(0.015 mole/liter), and $Na^+ : Ca^{2+} = 1 : 210$ in solution of sodium dodecylsulfate(0.006 mole/liter).

The air-solution interface of an aqueous solution of ionic surface active agent forms a uniformly charged surface owing to the adsorption of surface active ions, on which counterions are adsorbed from the solution. When the solution contains two kinds of counterions, the preferential adsorption is observed at the interface. The determination of the relative adsorbability of these two kinds of ions—*i.e.*, the ratio of the ratios of the two kinds of ions at the surface and in the solution—is very important in order (1) to know the distribution of ions at the charged surface, (2) to get insight into the selective permeability of counterions against ion-exchange resin membrane, and (3) to separate or concentrate the particular ions by foam fractionation. However, the determination of the concentration or composition of adsorbed counterions is difficult without the aid of the radiotracer technique. Judson *et al.* (3) have studied the relative adsorption between chloride and sulfate anions at the air-solution interface by radioactivity. Walling *et al.* (8) have found by foam fractionation that multivalent ions are preferentially adsorbed from a solution of sodium palmitoyl methyltaurine. As the concentrations of the solutions studied in these paper were above the c.m.c., a quantitative interpretation of the results was less accurate by the competitive adsorption of multivalent ions by micelles.

In the present investigation a large area of adsorbed surface was obtained by collecting well drained foams, and the concentration was determined by weighing. On the other hand, the amount of the adsorbed labeled ions was determined from the measurements of the ratio of counts of the dried samples of collapsed foams and of solution. The relative adsorbability was determined from these measurements.

Experimental

Materials. Radioactive $H^{36}Cl$, $K^{36}ClO_3$, and $H_2{}^{35}SO_4$ were obtained from the Daiichi Pure Chemicals Co. As the specific activity of these species was so high, each radioactive ion was diluted with each acid (or salt) solution. Dodecyltrimethylammonium bromide was synthesized from dodecylbromide (extra pure grade) and trimethylamine according to the procedures by Tartar (4). It was purified by recrystallization from acetone. The other salt—*e.g.*, dodecyltrimethylammonium nitrate—was obtained adding silver nitrate to solution of dodecyltrimethylammonium bromide. Dodecylammonium bromide, nitrate, chloride, and acetate were obtained by neutralizing dodecylamine with respective acids. Sodium sulfate and potassium bromide used were extra pure grade materials.

Procedures. Aqueous solutions of surfactant containing two kinds of counterions, one of which was labeled, were prepared. About 200 cc. of the solution of known concentration were introduced into an apparatus shown in Figure 1 (5). Narrow glass tubing was connected under the horizontal tubing to promote the circulation of the solution because of the movement of bubbles. The apparatus was kept in an air thermostat to carry out the experiment at constant temperature. The bubbles were generated by the action of a circulating pump. Bubbles attained adsorption equilibrium while they moved from one end to the other end of the nearly horizontal glass tube of about 50 cm. in length. The foams were well drained while they gradually moved upwards through a tube of about 60 cm. length. The tube was inclined to facilitate drainage of the foams. The foam collector was cooled to about 0°C. to collapse the foams. A correction because of the condensation of water vapor by cooling was necessary to determine the concentration of foams.

Fixed volumes of original solution and of collapsed foams were taken with a micro (overflow type) pipet and transferred to respective sample plates which were coated partly with silicone oil to prevent wetting. The solution was dried gradually with infrared ray lamp and then radioactive counts were determined. On the other hand, the concentration of the solution of collapsed bubbles was determined by weighing. The accuracy of this method was confirmed by drying the solution of known concentration at respective experiments. The loss of acetic acid (and therefore dodecylamine) was observed during the process of drying in the case of alkylamine acetate. Thus, the concentration of adsorbed molecules was determined in the form of dodecylammonium chloride by the addition of HCl prior to the drying. Relative adsorbability was determined from these values by the procedures described in the next section.

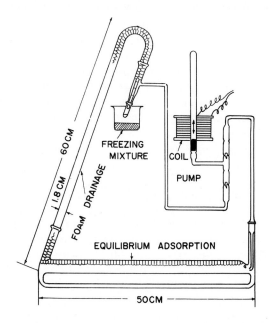

Figure 1. Apparatus of relative adsorption measurements

Results and Discussion

Several features of this experiment (5 ~ 7) as a radiotracer method are as follows: (1) since the relative adsorbability has been determined from the ratio of radiocounts of dried samples, it is not affected by errors in the absolute values for radioactivity and (2) specific activity, which has to be large usually for the study of surface phenomena, may be very small and yet the radioactive measurements yield accurate results. The concentrations of solutions examined were all close to but less than the c.m.c. values of respective surfactant solutions. The foams were stable enough to carry the experiment in this concentration range and the mole ratio of counterions in the bulk was equal to the stoichiometric mole ratio —*i.e.*, there was no shift in mole ratio of dispersed counterions owing to the micelle formation at the present study.

The adsorption of $^{36}Cl^-$ ions at the air-solution interface of an aqueous solution of dodecylammonium acetate (0.0140 mole/liter) containing a very small amount of chloride ions (0.000059 equiv./liter of $H^{36}Cl$) was determined, and the results were summarized in Table I. The concentration of the surfactant was maintained at an almost constant level below the c.m.c. throughout the experiment. An aliquot of the solution was dried and its radioactivity, R_s, determined is shown in the 1st line of Table I. The radioactivity of dried sample of the same volume of collapsed foams, R_f, is shown in the 2nd line. The concentration of surfactant (gram/liter) in solution is shown in the 3rd line and that of collapsed foams is determined by weighing and shown in the 4th line. It may be assumed that the composition of the interior of foams (film) is the same as that of the bulk solution, and the volume of adsorbed molecules can be neglected compared with the volume of collapsed foams. Thus, $C_f - C_s$ means the amount of adsorbed substances. The ratio of the fraction of labeled ions in the adsorbed state against that in solution is then given by,

$$\frac{(R_f - R_s)C_s}{R_s(C_f - C_s)} = \frac{X_{ads}}{X_{soln}} \quad (1)$$

and shown in the 5th line. Where X_{ads} and X_{soln} are the fractions of labeled ions among counterions in an adsorbed state and in solution. In the case when the fraction of labeled ions is very small compared with unity or the relative adsorbability is small, Equation 1 is very close to the relative adsorbability of labeled ions. The fraction of labeled ions among counterions in solution, X_{soln}, is given in the 6th line. The relative adsorbability (5) is then given by,

$$\alpha = \frac{X_{ads}}{1 - X_{ads}} \bigg/ \frac{X_{soln}}{1 - X_{soln}} \quad (2)$$

and shown in the 7th line of the Table.

Table I. Relative Adsorbability of Chloride Against Acetate Ions at the Air-Solution Interface of an Aqueous Solution of Dodecylammonium Acetate (0.0140 mole/liter) at 25°C.

	1	2	3
1. R_s (c.p.m.)	1387	1398	1363
2. R_f (c.p.m.)	8755	5780	12378
3. C_s (gram/liter)	3.43	3.43	3.43
4. C_f (gram/liter)	15.6	11.1	22.0
5. X_{ads}/X_{soln}	1.50	1.40	1.49
6. X_{soln}	0.0042	0.0042	0.0042
7. Relative adsorbability	1.49	1.40	1.49

The relative adsorbability of sulfate against bromide ions of aqueous solution of dodecyltrimethylammonium bromide (0.0150 mole/liter) containing a small amount of sulfate ions (0.0000594 equiv./liter of $Na_2{}^{35}SO_4$) was summarized in Table II.

Table II. Relative Adsorbability of $^{35}SO_4{}^{2-}$ Against Bromide at the Air-Solution Interface of an Aqueous Solution of Dodecyltrimethylammonium Bromide (0.0150 mole/liter) at 20°C.

	1	2	3	4
1. R_s (c.p.m.)	770	752	722	722
2. R_f (c.p.m.)	2850	2920	2470	2510
3. C_s (gram/liter)	4.64	4.69	4.62	4.62
4. C_f (gram/liter)	11.7	12.0	11.1	10.8
5. $X_{ads}/X_{soln} \approx \alpha$	1.75	1.84	1.72	1.86

The fraction of sulfate ions was 0.004, so that the ratio of mole fractions of sulfate ions in an adsorbed state and in solution was equal to the relative adsorbability within experimental error.

The relative adsorbability of chlorate against bromide ions of aqueous solution of dodecyltrimethylammonium bromide (0.0150 mole/liter) are summarized in Table III. The concentration of $K^{36}ClO_3$ was 0.00019 equiv./liter.

Table III. Relative Adsorbability of $^{36}ClO_3{}^-$ Against Br^- at the Air-Solution Interface of an Aqueous Solution of Dodecyltrimethylammonium Bromide at 25°C.

	1	2
1. R_s (c.p.m.)	2327	2318
2. R_f (c.p.m.)	15711	13493
3. C_s (gram/liter)	4.86	4.86
4. C_f (gram/liter)	19.34	17.31
5. X_{ads}/X_{soln}	2.07	2.02
6. X_{soln}	0.112	0.112
7. Relative adsorbability	2.40	2.32

As the fraction of chlorate ions among anions is not very small, the relative adsorbability differs appreciably from the ratio of the fractions of chlorate ions in an adsorbed state and in solution, X_{ads}/X_{soln}.

Similar experiments were carried out for various couples of counterions of aqueous solutions of dodecylammonium and dodecyltrimethylammonium salts. The average values of the relative adsorbability of various anions against chloride ions are summarized in Table IV.

Table IV. Relative Adsorbability of Anions Against Chloride

Surface active ions

Anions	$C_{12}H_{25}N^+(CH_3)_3$	c.m.c. (mole/liter)	$C_{12}H_{25}NH_3^+$	c.m.c.
CH_3COO^-	—	—	0.7 (25°C.)	0.015
Cl^-	1	0.020	1	0.014
Br^-	1.2 (20°C.)	0.017	1.5 (35°C.)	0.012
ClO_3^-	2.8 (25°C.)	—	3.0 (25°C.)	—
NO_3^-	1.6 (20°C.)	—	4.1 (25°C.)	0.0115
SO_4^-	2.3 (20°C.)	—	—	—

The foam stability of aqueous solution of dodecyltrimethylammonium acetate was not good enough to conduct the experiment. Added 1 : 1 type salts did not affect the relative adsorbability of univalent ions against univalent ions, but that of sulfate against bromide ions in solution containing only a small amount of sulfate ions changed with the concentration of added bromide as shown in Figure 2. The slope of log α vs. log C_{Br} was -1—i.e., the relative adsorbability of sulfate against bromide ions was inversely proportional to the concentration of bromide (see Figure 2).

The electrical potential, ϕ_0, of an ionized monolayer is given as a function of the concentration and valency of the counterions as follows (1),

$$\frac{2000\pi\sigma^2}{DNKT} = C_1 \exp\left(\frac{e\phi_0}{KT}\right) + C_2 \exp\left(\frac{2e\phi_0}{KT}\right) \tag{3}$$

where σ is the surface charge density, D the dielectric constant of the medium, e the elementary charge, C_i the concentration of i-valent counterions in mole/liter in solution, k the Boltzmann constant, and T the absolute temperature. The electrical energy decrease caused by the adsorption of a univalent counterion from the bulk to the ionized surface is $e\phi_0$, whereas that of bivalent counterion is $2e\phi_0$. It is evident from Equation 3 that the electrical potential, ϕ_0, decreases inversely proportional to log C_1 when C_2 is very small. It is expected that the difference of the electrical energy of adsorption between univalent and bivalent counterions decreases inversely proportional to log C_1 and the relative adsorbability of bivalent against univalent counterions changes inversely proportional to the concentration of univalent counterions. The agreement between this conclusion and experimental results shown in Figure 2 suggests the electrical energy is an important factor at the adsorption of counterions in $C_{12}H_{25}N(CH_3)_3Br$—SO_4 system as well as in $C_{12}H_{25}SO_4$–Na—Ca system (6). But, it seems difficult at the present stage, to explain the great difference between the relative adsorbability, 1.9, of sulfate against bromide ions and that, 210, of calcium against sodium ions, in terms of the differences in hydration energy, ionic radius, etc. A very large value for $PoCl_6^-$ against Cl^- was obtained by Ter-Minassian-Saraga *et al.* (2). Experimental data, such as the c.m.c. values and/or selective permeability of anions against ion-exchange resin membrane, are in parallel with the relative adsorbability of respective ions.

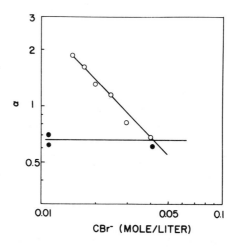

Figure 2. Dependence of relative adsorbability, α, on the concentration of bromide, C_{Br^-}

○; SO_4^{2-}: Br^- in $C_{12}H_{25}N^+(CH_3)_3$ (20°C.)
●; Cl^-: Br^- in $C_{12}H_{25}NH_3^+$ (30°C.)

Summary

A large adsorbed surface phase was accumulated by collecting well drained foams. The amount of adsorbed species was determined by weighing the dried collapsed foams. The amount of labeled counterion adsorbed was determined from the measurements of the ratio of the radioactive counts from the known volumes of collapsed foams and the bulk solution. Thus, the ratio of two kinds of adsorbed counterions at the surface was obtained. The relative adsorbability of counterions was determined from the ratio of mole ratios in an adsorbed and in the bulk phases. The distribution of counterions at the charged surface and in the bulk solution was determined, combining the radiotracer technique with foam fractionation technique. The relative adsorbability was Cl^- : CH_3COO^- : Br^- : ClO_3^- : NO_3^- = 1 : 0.7 : 1.5 : 3.0 : 4.1 in solution of dodecylammonium (0.011 mole/liter), Cl^- : Br^- : ClO_3^- : NO_3^- : SO_4^{2-} = 1 : 1.2 : 2.8 : 1.6 : 2.3 in solution of dodecyltrimethylammonium (0.015 mole/liter), and Na^+ : Ca^{2+} = 1 : 210 in solution of sodium dodecylsulfate (0.006 mole/liter).

Literature Cited

(1) Cassie, A. B. D., Palmer, R. C., *Trans. Faraday Soc.* **37,** 156 (1941).
(2) Hendrikx, Y., Luzzati, A., Ter-Minassian-Saraga, L., *J. Chim. Phys.* **61,** 1351 (1964).
(3) Judson, C. M., Lerew, A. A., Dixon, J. K., Salley, D. J., *J. Phys. Chem.* **57,** 916 (1953).
(4) Scott, A. B., Tartar, H. V., *J. Am. Chem. Soc.* **65,** 693 (1943).
(5) Shinoda, K., Mashio, K., *J. Phys. Chem.* **64,** 54 (1960).
(6) Shinoda, K., Ito, K., *J. Phys. Chem.* **65,** 1499 (1961).
(7) Shinoda, K., Nakanishi, J., *J. Phys. Chem.* **67,** 2547 (1963).
(8) Walling, C., Ruff, E. E., Thornton, J. L. Jr., *J. Phys. Chem.* **61,** 486 (1957).

RECEIVED October 26, 1967.

INDEX

INDEX

A

Acid adsorption 127
Active carbon, reaction of hydrated
  proton with 112
Adsorbability of counterions at the
  charged interface, relative ... 198
Adsorbent, glass powder as 47
Adsorbing site, hydrophobic character of the 27
Adsorption
  alkali metal ion 78
  vs. alkyl chain length 165
  of anions 82
  of aqueous Co(II) at the silica-
    water interface 62
  of aqueous metal on colloidal
    hydrous manganese oxide .. 74
  on carbon 112
  cobalt ion 78
  copper ion 78
  of dyes 173
  envelope 85
  equation, Gibbs 5
  equilibria 113
  global 9
  of hafnium species on silver
    iodide sols 60
  of hydrocarbon gases on silica gel 13
  of hydrolyzable ions 46
  of hydrolyzed hafnium ions on
    glass 44
  of hydronium ion 112
  isotherms 83, 149, 165, 176, 189
    for spherical particles 151
  kinetics of 8, 25
  mechanism of 85
  negative 146
  nickel ion 78
  pH vs. 76
  phenomena of Graphon in aqueous surfactant solutions ... 135
  polylayer 189
  of pseudocyanine, energy of ... 193
  rates of 113
  reversibility of 83
  of selenite by geothite 82
  specific 76
  of strong electrolyte 145
  of sulfonates at solid/water interface, effect of hydrocarbon
    chain length on 161
Affinity cuts 35
Agglomeration of silica 101
Air-solution interface 198

Alkali metal ion adsorption 78
Alkylcarbocyanines, meso- 194
Alkyl chain length vs. adsorption .. 165
9-Alkylthiocarbocyanines 195
Alumina/water interface 162
Aluminum(III), coagulation by .. 91
Aluminum(III), hydrolysis of ... 97
Anions, adsorption of 82
Aqueous metal on colloidal hydrous
  manganese oxide, adsorption of 74
Aqueous surfactant solutions, adsorption and wetting phenomena of Graphon in 135
Area of β-naphthalene sulfonate
  ion, limiting 158
Astraphloxin 175
Axial dispersion 10

B

Benzpyran 113
B.E.T. 75, 155, 162
Breakthrough curves 16

C

Capillary thermodynamics 1
Carbon, adsorption on 112
Carbon, reaction of hydrated proton
  with active 112
Charged interface, relative adsorbability of counterions at the .. 198
Charged sites, separation between
  the 27
Chemical model 20
Chemisorption 15
Chromatographic behavior of interfering solutes 30
Chromatographic techniques 20
Chromene 113
  -acid reaction 131
Coagulation 136
  by aluminum(III) 91
  kinetics of 91
  orthokinetic 94
  perikinetic 93
  of silica by Co(II) 71
Cobalt ion adsorption 78
Co(II), coagulation of silica by .. 71
Co(II) at the silica-water interface,
  adsorption of aqueous 62
Coefficient of selectivity 26
Coherence 33
Colloidal hydrous manganese oxide,
  adsorption of aqueous metal on 74

Colloids, types of	45
Complex ion, cross-sectional area of hydrolyzed	47
Composite isotherm	145
Composition paths	35
Composition velocity	36
Concentration velocities	32
Constant potential isotherm	157
Contact angles	138
Copper ion adsorption	78
Counterion in diffuse double layer, sodium ion as	159
Counterions	166
at the charged interface, relative adsorbability of	198
Cross-sectional area of Hf(OH)$_4$ complex	59
Cross-sectional area of hydrolyzed complex ion	47
Cyanine dyes	173

D

Debye-Hückel approximation	154
Debye-Hückel law, extended	114
Diffuse double layer	145
Guoy-Chapman model of	150
sodium ion as counterion in	159
Diffusion coefficients	118
Diffusion equations	115
Diffusivity, internal	11
Dispersibility	136
Dispersion, axial	10
Dispersion of fine particles	135
Distance-time diagram	40
Dodecyltrimethylammonium bromide	24, 137, 199
Double layer	145
electrical	162
Guoy-Chapman model of diffuse	150
thickness vs. depth of surface region, electrical	155
DTAB	137
Dyes, adsorption and surface spectra of	173
Dyes, surface concentrations of	194

E

Energy of adsorption of pseudocyanine	193
Enthalpy of exchange, free	26
Entropy, excess	3
Envelope, adsorption	85
Electrical double layer	145, 162
thickness vs. depth of surface region	155
Electric charges	17
Electric gradient	16
Electric potential	21
Electrolyte, adsorption of strong	145
Electrophoresis	42
Electrophoretic mobility	163
of quartz	65
Equilibria, adsorption	113
Excess entropy	3

Excess surface charge of geothite	84
Exchange free enthalpy of	26
isotherm	83
reaction of	26
External transfer	9

F

Films, soap	25
Fine particles, dispersion of	135
Foam fractionation	198
Foams, collapsed	199
Free enthalpy of exchange	26
Freundlich equation	117
Freundlich parameters	124, 128, 131

G

Geothite, adsorption of selenite by	82
Geothite, excess surface charge of	84
Gibbbs adsorption equation	5
dividing surface	1
-Duhem equation	2
surface excess	149
thermodynamic analysis of fluid systems	1
Glass, adsorption of hydrolyzed hafnium ions on	44
Glass powder as adsorbent	47
Global adsorption	9
Graphetized carbon black	146
Graphon in aqueous surfactant solutions, adsorption and wetting phenomena of	135
Guggenheim-Adam "N" convention surface excess	149
Guoy-Chapman expression	70
Guoy-Chapman model of diffuse double layer	150

H

Hafnium hydroxide, precipitation of	48
Hafnium ions on glass, adsorption of hydrolyzed	44
Hafnium species, hydrolyzed	52
Hafnium species on silver iodide sols, adsorption of	60
Halide surface areas	194
Hemimicelle	167
Hexadecyldimethylethylolammonium bromide	24
Hexadecyltrimethylammonium bromide	24
Hf(OH)$_4$ complex, cross-sectional area of	59
Hf(OH)$_4$, surface coverage by	59
H-function	39
h-transformation	40
Hydrated proton with active carbon, reaction of	112
Hydrocarbon chain length on adsorption of sulfonates at solid/water interface, effect of	161

INDEX

Hydrocarbon gases on silica gel, adsorption of 13
Hydrogen ion activity 118
Hydrolysis of aluminum(III) 97
Hydrolyzable ions, adsorption of .. 46
Hydrolyzed
  aluminum 91
  complex ion, cross-sectional area of 47
  hafnium species 52
  polyvalent metal ions 95
Hydronium ion, adsorption of ... 112
Hydronium ion adsorption, salt effects on 123
Hydrophobic character of the absorbing site 27

I

Interface
  adsorption of aqueous Co(II) at the silica-water 62
  air-solution 198
  alumina/water 162
  effect of hydrocarbon chain length on adsorption of sulfonates at solid/water 161
  relative adsorbability of counterions at the charged 198
Interfering solutes, chromatographic behavior of 30
Internal diffusion 15
Internal diffusivity 11
Internal transfer 10
Intraparticle transport 115
Ion exchange 166
  processes 16
Ion exchangers, monomolecular .. 23
Ionized monolayer 150
Ion selectivity 27
Isotherm, composite 145
Isotherm, constant potential 157

K

Kinetics of adsorption 8, 25
Kinetics of coagulation 91
Kubelka-Munk reflectivity function 183

L

Langmuir adsorption law 181
Limiting area of β-naphthalene sulfonate ion 158

M

Manganese nodules 75
Manganese oxide, adsorption of aqueous metal on colloidal hydrous 74
Mechanism of adsorption 85
Meso-alkylcarbocyanines 194
Metal on colloidal hydrous manganese oxide, adsorption of aqueous 74

Metal ions
  on glass, adsorption of 44
  hydrolyzed polyvalent 95
  from solution, adsorption of ... 62
Mica 194
  synthetic 175
Mobility of silver iodide sol 60
Monolayer 189
  ionized 150
Monomolecular ion exchanges 23
Multilayers 189

N

β-Naphthalene sulfonate ion, limiting area of 158
β-Naphthalene sulfonate, sodium . 146
Negative adsorption 146
Nernst-Planck 16
Nickel ion adsorption 78
Noncoherent profiles 36

O

One component systems 2
Orthokinetic coagulation 94

P

Particle velocity 32
Perchlorate ion 126
Perikinetic coagulation 93
pH vs. adsorption 76
$PoCl_6^{2-}$ ions 25
Point coefficient 12
Point processes 9
Point resistances 19
Poisson-Boltzmann equation 150
Polylayer adsorption 189
Polyvalent metal ions, hydrolyzed 95
Precipitation of hafnium hydroxide 48
Profiles, noncoherent 36
Proton with active carbon, reaction of hydrated 112
Pseudo-colloids 45
Pseudocyanine 173
  energy of adsorption of 193

Q

α-Quartz 63
Quartz, electrophoretic mobility of 65
Quartz, turbidity of 66

R

Radiocolloids 45
Radiotracer technique 198
Reaction of exchange 26
Reflectivity function, Kubelka-Munk 183
Relative adsorbability of counterions at the charged interface 198
Resistances, space and point 19
Reversibility of adsorption 83
Root velocity 40

S

Salt effects of hydronium ion adsorption	123
SDS	137
Selectivity, coefficient of	26
Selenite by geothite, adsorption of	82
Silica	
agglomeration of	101
by Co(II), coagulation of	71
colloids, stability of	106
gel, adsorption of hydrocarbon gases on	13
water interface, adsorption of aqueous Co(II) at the	62
Similions	145
Silver bromide dispersions	175
Silver, colloidal	175
Silver iodide sol, mobility of	60
Silver iodide sols, adsorption of hafnium species on	60
Soap films	25
Sodium dodecyl sulfate	137
Sodium ion as counterion in diffuse double layer	159
Sodium β-naphthalene sulfonate	146
Solid/water interface, effect of hydrocarbon chain length on adsorption of sulfonates at	161
Soluble species of $Hf(OH)_4$	46
Solutes, chromatographic behavior of interfering	30
Space	
diffusion	21
processes	9
resistances	19
velocities	32
Specific adsorption	76
Spherical particles, adsorption isotherms for	151
Spheron 6	138
Stability of silica colloids	106
Stern-Grahame treatment	70, 169
Stern plane	162
Sterling FTG	154
Strong electrolyte, adsorption of	145
Sulfonates at solid/water interface, effect on hydrocarbon chain length on adsorption of	161
Surfactant solutions, adsorption and wetting phenomena of Graphon in aqueous	135
Surface charge density	154
Surface concentrations of dyes	173, 194
Surface coverage by $Hf(OH)_4$	59
Surface excess, Gibbs	149
Surface excess, Guggenheim-Adam "N" convention	149
Surface potential	99, 150
maximum	157
Surface region	
electrical double layer thickness $vs.$ depth of	155
thickness of	151
Surface spectra of dyes	173

T

Thermodynamics, capillary	1
Thiatricarbocyanine	195
Thickness of surface region	151
Traffic jam	32
Trajectories	40
Transfer, external and internal	9
True colloids	45
Turbidity of quartz	66
Two component systems	4

V

Velocities, space and concentration	32
Velocities, particle and wave	32
Velocity, composition	36
Velocity, root	40
Volta potential	146

W

Watershed point	35
Water/solid interface, effect of hydrocarbon chain length on adsorption of sulfonates at	161
Wave velocity	32
Wetting phenomena of Graphon in aqueous surfactant solutions	135

Z

Zero-point-of-charge	75, 163